Spin Physics

Selected Papers from
the 21st International Symposium on Spin Physics
(SPIN2014)

Peking University, Beijing, China, 20–24 October 2014

Editors

Haiyan Gao
Duke University, USA

Bo-Qiang Ma
Peking University, China

 **World Scientific**

NEW JERSEY · LONDON · SINGAPORE · BEIJING · SHANGHAI · HONG KONG · TAIPEI · CHENNAI · TOKYO

Published by

World Scientific Publishing Co. Pte. Ltd.
5 Toh Tuck Link, Singapore 596224
USA office: 27 Warren Street, Suite 401-402, Hackensack, NJ 07601
UK office: 57 Shelton Street, Covent Garden, London WC2H 9HE

British Library Cataloguing-in-Publication Data
A catalogue record for this book is available from the British Library.

SPIN PHYSICS
Selected Papers from The 21st International Symposium on Spin Physics (SPIN2014)

ISBN 978-981-3142-70-1
ISBN 978-981-3109-95-7 (pbk)

Typeset by Stallion Press
Email: enquiries@stallionpress.com

Printed in Singapore

Spin Physics

CONTENTS

Memorial Session

Spin Physics (SPIN2014)
International Journal of Modern Physics: Conference Series
Vol. 40 (2016) 1602001 (7 pages)
© World Scientific Publishing Company
DOI: 10.1142/S2010194516020018

Foreword

Published 29 February 2016

The 21st International Symposium on Spin Physics (SPIN2014) was successfully held on the campus of Peking University, Beijing, China, during October 20–24, 2014. The scientific program of this symposium included many topics related to spin phenomena in particle and nuclear physics as well as those in related fields. The International Spin Physics Symposium series has combined the High Energy Spin Symposia and the Nuclear Polarization Conferences since 2000. The most recent symposia were held in JINR, Dubna, Russia (2012), in Forschungszentrum Jülich, Germany (2010), and in Charlottesville, Virginia, USA (2008).

The symposium was hosted by Peking University, and supported by many renowned research institutions and universities both inside and outside China. The scientific program consisted of plenary sessions and parallel sessions including the following topics:

- Spin structure of hadrons
- Spin structure of nucleon (longitudinal)
- Spin Structure of nucleon (transverse)
- Nucleon structure (form factors, GPDs)
- Spin in hadronic reactions
- Spin physics with photons and leptons
- Spin physics in nuclear reactions and nuclei
- Fundamental symmetries and spin physics beyond the Standard Model
- Accelerator, storage and polarimetry of polarized beams
- Polarized ion and lepton sources and targets
- Future facilities and experiments
- Applications of spin physics

In addition, there were a memorial session in honor of Professor Michel Borghini and a public lecture presented by Professor Qi-Kun Xue at Tsinghua University during the symposium.

The current proceedings comprise written contributions of many presentations during SPIN2014.

Foreword

We gratefully acknowledge the supports from the members of the International Spin Physics Committee, the parallel program conveners and all the other organizers and participants. We acknowledge the local organizers and especially the following students: Xianda Liu, Tianbo Liu, Xiaozhen Du, Wenjuan Mao, Zhentao Zhang, Lijing Shao, Jun Zhang, Bofeng Wu, Yanqi Huang *et al.* for their contributions towards the successful organization of this symposium.

Haiyan Gao & Bo-Qiang Ma

Committees of the 21th International Symposium on Spin Physics (SPIN2014)

International Spin Physics Committee:

R. Milner — MIT (Chair)
E. Steffens — Erlangen (Past-Chair)
M. Anselmino — Torino
E. Aschenauer — BNL
A. Belov — INR Moscow
H. Gao — Duke
P. Lenisa — Ferrara
B.-Q. Ma — Peking
N. Makins — Illinois
A. Martin — Trieste
A. Milstein — Novosibirsk
M. Poelker — JLab
R. Prepost — Wisconsin
T. Roser — BNL
N. Saito — KEK
H. Sakai — Tokyo
H. Stroeher — Juelich
O. Teryaev — Dubna
F. Bradamante* — Trieste
E.D. Courant* — BNL
D.G. Crabb* — Virginia
A.V. Efremov* — JINR
G. Fidecaro* — CERN
W. Haeberli* — Wisconsin
A.D. Krisch* — Michigan
A. Masaike* — Kyoto
C.Y. Prescott* — SLAC
V. Soergel* — Heidelberg
W.T.H. van Oers* — Manitoba
(* honorary member)

Symposium Co-chairs:

Haiyan Gao — Duke University   Bo-Qiang Ma — Peking University

Local Organization Committee:

Xiangsong Chen (Huazhong Univ. of Science and Tech.)
Haiyan Gao (Duke, Co-Chair)
Bitao Hu (Lanzhou Univ.)
Xiaomei Li (China Inst. of Atomic Energy (CIAE))
Zuotang Liang (Shandong Univ.)
Chuan Liu (Peking Univ.)
Haijing Lv (Huangshan Univ.)
Bo-Qiang Ma (Peking Univ., Co-Chair)
Jianping Ma (Inst. of Theo. Phys. (ITP), CAS)
Yajun Mao (Peking Univ.)
Zhigang Xiao (Tsinghua Univ.)
Bowen Xiao (Central China Normal U)
Qinghua Xu (Shandong Univ.)
Wenbiao Yan (Univ. of Science and Tech. of China (USTC))
Qiang Zhao (Inst. of High Energy Physics (IHEP), CAS)
Shi-Lin Zhu (Peking Univ.)
Bingsong Zou (Inst. of Theo. Phy., CAS)
Chao-Hsi Chang* (ITP, CAS, Beijing)
Kuang-Ta Chao* (Peking Univ., Beijing)
H. Y. Cheng* (Academia Sinica, Taipei)
Hideto En'yo* (RIKEN, Wako)
Hanxin He* (CIAE, Beijing)
Tao Huang* (IHEP, CAS, Beijing)
Xiangdong Ji* (Univ. of Maryland/Shanghai Jiao Tong Univ.,
            College Park/Shanghai)
Anthony W. Thomas* (Univ. of Adelaide, Adelaide)
Fan Wang* (Nanjing Univ., Nanjing)
Zhengguo Zhao* (USTC, Hefei)
(* advisor to LOC)

Parallel Program Conveners:

S1: Spin Structure of Hadrons (joint)
        — Chuan Liu, Vincenzo Barone, Feng Yuan, Akio Ogawa
S2: Spin Structure of Nucleon (longitudinal)
        — Renee Fatemi, Jian-Ping Chen, Marco Stratmann
S3: Spin Structure of Nucleon (transverse)
        — Alexei Prokudin, Zuo-Tang Liang, Andrea Bressan
S4: Nucleon Structure (form factors, GPDs)
        — Nicole D'Hose, John Arrington, Dieter Mueller, Barbara Pasquini
S5: Spin in Hadronic Reactions
        — Seonho Choi, Frank Rathmann

S6: Spin Physics with Photons and Leptons
        — Michael Ostrick, Reinhard Beck, Mohammad Ahmed
S7: Spin Physics in Nuclear Reactions and Nuclei
        — Arnoldas Deltuva, Evgeny Epelbaum, Kimiko Sekiguchi, Zhigang Xiao,
S8: Fundamental Symmetries and Spin Physics Beyond the Standard Model
        — Krishna Kumar, Jens Erler, Yuri Obukhov
S9: Accelerator, Storage and Polarimetry of Polarized Beams
        — Alexander Chao, Mei Bai, Andreas Lehrach
S10: Polarized Ion and Lepton Sources and Targets
        — Matt Poelker, Don Crabb
S11: Future Facilities and Experiments
        — Rolf Ent, Jianwei Qiu, Zhengguo Zhao, Alexander Nagaytsev
S12: Applications of Spin Physics
        — Gordon Cates, Warren Warren (medical),
        — Guilu Long and Xincheng Xie (quantum info),
        — Xi Chen and Ruirui Du (CMP)
Public Lecture — Zhigang Xiao, Haiyan Gao
Borghini Memorial Session — Alan Krisch, Akira Masaike
Poster Session — Xiaomei Li

Sponsoring Institutions:

School of Physics, Peking University
State Key Laboratory of Nuclear Physics and Technology, Peking University
Center for High Energy Physics, Peking University
Duke University
Brookhaven National Laboratory
Central China Normal University
European Physical Journal A
Jefferson Lab
Huangshan University
Huazhong University of Science and Technology
Institute of High-Energy Physics
Institute of Theoretical Physics
Lanzhou University
Shandong University
Tsinghua University
University of Science and Technology of China
National Natural Science Foundation of China

The websites of Spin2014: http://www.phy.pku.edu.cn/spin2014/
http://indico.cern.ch/event/284740/

Conference site of SPIN2014

Conference photo of SPIN2014

Plenary Session

Spin Physics (SPIN2014)
International Journal of Modern Physics: Conference Series
Vol. 40 (2016) 1660001 (10 pages)
© The Author(s)
DOI: 10.1142/S2010194516600016

The Spin Structure of the Nucleon

Xiangdong Ji

*INPAC, Department of Physics and Astronomy, Shanghai Jiao Tong University,
Shanghai, 200240, P. R. China*

*Center for High-Energy Physics, Peking University,
Beijing, 100080, P. R. China*

*Maryland Center for Fundamental Physics, University of Maryland,
College Park, MD 20742, USA
xji@physics.umd.edu*

Yong Zhao

*Maryland Center for Fundamental Physics, University of Maryland,
College Park, MD 20742, USA
yongzhao@umd.edu*

Published 29 February 2016

We justify the physical meaning of the spin and orbital angular momentum of free
partons in the infinite momentum frame, and discuss the relationship between the Jaffe-
Manohar and Ji's sum rules for proton spin. The parton orbital angular momentum in
the Jaffe-Manohar sum rule can be measured through twist-three GPD's in hard scattering processes such as deeply virtual Compton scattering. Furthermore, we propose that
the paton orbital angular momentum as well as the gluon helicity can be calculated in
lattice QCD through a large momentum effective theory approach, and provide all the
one-loop matching conditions for the proton spin content in perturbative QCD.

Keywords: Proton spin sum rule; Twist-three GPD; Lattice QCD.

PACS numbers: 14.20.Dh, 13.88.+e, 12.38.Gc, 12.38.Bx

1. Introduction

It is an important goal in hadron physics to understand the spin structure of the
nucleon. In 1987, the European Muon Collaboration (EMC) at CERN discovered
that the quark spin only accounts for a very small portion of the longitudinal proton
spin,[1,2] which is far less than expected from people's understanding at that time.
Since then, the EMC result has inspired generations of hadron physics programs

to study the spin content of the proton. Among them are SLAC, SMC and COMPASS at CERN, HERMES at DESY, PHENIX, STAR and BRAHMS at RHIC, and JLab (see a recent review in Ref. 3). Our current understanding of the proton spin structure is that the quark spin contributes about one third,[4] while there is strong evidence that the gluon polarization also shares a considerable positive amount;[5] the rest is believed to be distributed between the quark and gluon orbital angular momentum (OAM). With the JLab 12GeV upgrade[6] as well as the electron-ion collider (EIC),[7] people will be able to obtain more abundant and accurate information about the structure of proton spin.

The key issues associated with the proton spin structure are:

- What is a physical sum rule for the proton spin?
- How to probe the OAM of the quark and gluon partons?
- How to calculate the gluon helicity and parton OAM in lattice QCD?

2. The Jaffe-Manohar and Ji's Sum Rules for Proton Spin

In the past 25 years, two well-known spin sum rules have been proposed to analyze the proton spin structure. The first, proposed by Jaffe and Manohar,[8] was motivated from a free-field expression of QCD angular momentum boosted to the infinite momentum frame (IMF) of the proton. The second, usually called Ji's sum rule, is the frame-independent and manifestly gauge-invariant decomposition by one of the authors.[9]

The Jaffe-Manohar sum rule is defined in the light-cone gauge $A^+ = 0$, and states that the proton spin can be decomposed into four parts,

$$\frac{1}{2} = \frac{1}{2}\Delta\Sigma(\mu) + l_q^z(\mu) + \Delta G(\mu) + l_g^z(\mu), \tag{1}$$

where the individual terms are the spin and OAM of the quark and gluon partons, respectively, and μ is a renormalization scale. All the four terms are defined to be the proton matrix elements of free-field angular momentum operators in the IMF:[8]

$$\vec{J} = \int d^3x \ \psi^\dagger \frac{\vec{\Sigma}}{2}\psi + \int d^3x \ \psi^\dagger \vec{x} \times (-i\vec{\nabla})\psi$$

$$+ \int d^3x \ \vec{E}_a \times \vec{A}^a + \int d^3x \ E_a^i \ \vec{x} \times \vec{\nabla}A^{i,a}, \tag{2}$$

where $E^i = F^{i+}$, a and i are the color and spatial indices. Here the system is fixed on the light-cone plane, i.e., $x^+ = (x^0 + x^3)/\sqrt{2} = 0$, which is equivalent to the IMF.

In the light-cone gauge, each term in Eq. (2) can be expressed as sum of the spin and OAM over all Fock states, so the Jaffe-Manohar sum rule has a clear partonic interpretation. However, the free-field form of the angular momentum in gauge theories faces two conceptual problems: all terms except the first one are gauge dependent, and it is unclear why the light-cone gauge operator is measurable in experiments.

Ji's sum rule takes a different form from Eq. (2), as the total QCD angular momentum is decomposed into three gauge-invariant parts:[9]

$$\vec{J} = \int d^3x \ \psi^\dagger \frac{\vec{\Sigma}}{2} \psi + \int d^3x \ \psi^\dagger \vec{x} \times (-i\vec{\nabla} - g\vec{A})\psi$$
$$+ \int d^3x \ \vec{x} \times (\vec{E} \times \vec{B}), \tag{3}$$

where the total gluon angular momentum cannot be gauge-invariantly decomposed into its spin and OAM. In this way, Ji's sum rule reads:

$$\frac{1}{2} = \frac{1}{2}\Delta\Sigma(\mu) + L_q^z(\mu) + J_g^z(\mu) \ . \tag{4}$$

Notwithstanding that Ji's sum rule has received considerable attention for its relation to generalized parton distributions (GPD's) and experimental probes,[9–11] it is not natural in the language of parton physics (see, however, a recent discussion on its connection to the transverse polarization[12]).

Since the gluon polarization is not defined from the Ji sum rule, it has generated a lot of theoretical attempts to define gauge-invariant gluon spin and OAM angular momentum operators.[13–16] As has been discussed extensively in a recent review,[17] these new sum rules can be classified into two classes—the Jaffe-Manohar and Ji types. There are two key issues that distinguish these two types of sum rules: the Jaffe-Manohar sum rule is frame dependent as it is defined in the IMF, but the Ji's sum rule does not depend on any frame. Besides, as one can compare Eq. (2) and Eq. (3), the Jaffe-Manohar sum rule uses the canonical OAM operators, while Ji's sum rule adopts the "mechanical" OAM operator that results from the Belinfante-Rosenfeld procedure.

In the past two decades, there has been a long list of literatures attempting to justify the Jaffe-Manohar sum rule as physical (see e.g., Refs. 18, 19, 20, 21, 22, 23, 24). There are strong motivations behind this: First, ΔG as defined in the light-cone gauge is measurable in high-energy experiments, although this appears to be a theoretical puzzle by itself—while ΔG is easy to define from the Feynman parton picture, there is no natural gauge-invariant notion for the spin of gauge particles.[25] Second, the simple partonic picture of the Jaffe-Manohar sum rule in the IMF makes it more natural to use free-field operators.

It was first proposed that although the free-field angular momentum operators are gauge dependent, their physical matrix elements are gauge invariant.[19] A similar claim was also made recently.[20] However, this is invalidated by a one-loop calculation,[23] where the matrix element of the gluon spin operator was shown to be different in the Coulomb and light-cone gauges, as has been proved in Ref. 26 (for more general discussions see Ref. 21). Actually, as argued in Refs.,[22,23] for the bound-state proton, there is no physically meaningful notion of gluon spin or OAM due to the existence of longitudinal gluons. Only when the proton is boosted to IMF, the longitudinal component of gluons is suppressed by the infinite boost and the gluons can be regarded as free radiation. This is the well-known Weizsäcker-Williams (WW) approximation.[27] The gluon spin then acquires a clear physical meaning and can be represented by $\vec{E} \times \vec{A}$, but is subject to a class of "physical" gauge conditions that leave the transverse polarizations of the gluon field intact.[24] Similar arguments also apply to the quark and gluon OAM. Therefore, we can regard the free-field form in the Jaffe-Manohar sum rule as physical if we work in IMF with a "physical" gauge condition. This is equivalent to using the light-cone coordinates and gauge,[8] and the reason is simple: All the "physical" gauges will flow into the light-cone gauge in the IMF limit.[24]

3. Probing the Parton Orbital Angular Momentum

When Ji's sum rule was first proposed, it immediately receives a lot of attention because each term can be measured through twist-two GPD's from deeply virtual Compton scattering (DVCS) experiments.[9, 10] In Ji's sum rule, the total quark and gluon angular momenta satisfy

$$J_{q,g} = \frac{1}{2} \left[A_{q,g}(0) + B_{q,g}(0) \right],$$

$$J_q + J_g = \frac{1}{2}, \tag{5}$$

where $A_{q,g}(0)$ and $B_{q,g}(0)$ are form factors of the symmetrized quark and gluon energy-momentum tensors.

In a DVCS process as shown in Fig. 1, the Compton amplitude depends on four twist-two GPD's, H, $\tilde{H}$, E and $\tilde{E}$. In the light-cone gauge, they are defined to be the off-forward matrix elements of the light-cone correlations:

$$\int \frac{d\lambda}{2\pi} e^{i\lambda x} \langle P' | \bar{\psi}(-\frac{\lambda n}{2}) \gamma^\mu \psi(\frac{\lambda n}{2}) | P \rangle = H(x, \Delta^2, \xi) \bar{U}(P') \gamma^\mu U(P)$$

$$+ E(x, \Delta^2, \xi) \bar{U}(P') i \frac{\sigma^{\mu\nu} \Delta_\nu}{2M} U(P) + \cdots,$$

$$\int \frac{d\lambda}{2\pi} e^{i\lambda x} \langle P' | \bar{\psi}(-\frac{\lambda n}{2}) \gamma^\mu \gamma_5 \psi(\frac{\lambda n}{2}) | P \rangle = \tilde{H}(x, \Delta^2, \xi) \bar{U}(P') \gamma^\mu U(P)$$

$$+ \tilde{E}(x, \Delta^2, \xi) \bar{U}(P') i \frac{\sigma^{\mu\nu} \Delta_\nu}{2M} U(P) + \cdots, \tag{6}$$

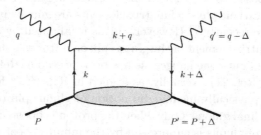

Fig. 1. Dominant scattering process in DVCS.

where $|P\rangle$ is a plane-wave proton state with momentum $P^\mu = (P^0, 0, 0, P)$, n^μ is a vector along the light-cone direction, and the skewness parameter $\xi = -n \cdot \Delta / n \cdot (P + P')$. Here the "$\cdots$" represents higher-twist contributions. The proton spin is related to these GPD's through the sum rule[9, 10]

$$\int_{-1}^{1} dx \; x \left[H(x, \xi, \Delta^2) + E(x, \xi, \Delta^2) \right] = A_q(\Delta^2) + B_q(\Delta^2) . \tag{7}$$

To obtain J_q, one can extrapolate the sum rule to $\Delta^2 = 0$, and then $J_g = 1/2 - J_q$. Since the quark spin $\Delta\Sigma$ has been precisely measured in inclusive and semi-inclusive scattering experiments, one can subtract it from J_q to determine L_q^z in Ji's sum rule. In the IMF, J_g can be further decomposed into three parts,

$$J_g^z = \Delta G + l_g^z + J_{\text{pot}}^z, \tag{8}$$

where the so called "potential" angular momentum J_{pot}^z corresponds to the operator

$$\vec{J}_{\text{pot}} = g \int d^3 x \; \psi^\dagger \vec{x} \times \vec{A} \psi . \tag{9}$$

Although ΔG is known to be measurable, to determine l_g^z in the Jaffe-Manohar sum rule one still needs to measure the contribution from J_{pot}^z, which is also the case for l_q^z.

Since the parton OAM involves the transverse motion of the patrons, it is intrinsically a twist-three observable. However, for L_q^z in Ji's sum rule, it can be related to twist-two GPD's through rotational symmetry, because the operator is frame independent. In contrast, l_q^z and l_g^z in the Jaffe-Manohar sum rule do not have frame independence, and to measure them as well as J_{pot}^z one has to rely on twist-three GPD's.[28, 30] The relationships between l_q^z, l_g^z and J_{pot}^z are derived in Ref. 30, and we are not going to provide the details here. Meanwhile, these twist-three GPD's have been studied and can be extracted from two-photon processes such as DVCS.[31, 32]

4. Lattice Calculation of the Proton Spin Content

Although we have justified its physical meaning, the Jaffe-Manohar sum rule itself still poses difficulty for a nonperturbative lattice calculation of its individual contributions, because the explicit usage of light-cone coordinates and gauge brings real-time dependence. One may avoid this difficulty by using normal space-time coordinates with a "physical" gauge that does not involve time, and calculating with a proton at infinite momentum. However, the largest momentum attainable on the lattice with spacing a is constrained by the lattice cutoff π/a.

Ji's sum rule, instead, is frame independent so that one can go off the light-cone and work in a reference frame with finite momentum. Lorentz covariance guarantees that each component in this sum rule will be the same in the IMF. Therefore, a considerable amount of lattice QCD calculations have been carried out to study

Ji's sum rule, among which are the LHPC,[33,34] SESAM,[33] USQCDSF-UKQCD,[35] ETMC[36,37] and χQCD[38] collaborations.

The difficulty in the lattice study of the Jaffe-Manohar sum rule can, however, be circumvented in the framework of *large-momentum effective field theory* (LaMET)[39] proposed by one of the authors. Suppose one is to calculate some light-cone or parton observable $\mathcal{O}$. Instead of computing it directly, one defines, in the LaMET framework, a quasi-observable $\tilde{\mathcal{O}}$ that depends on a large hadron momentum P^z. In general, both $\mathcal{O}$ and $\tilde{\mathcal{O}}$ suffer from ultraviolet (UV) divergences. If $P^z \to \infty$ is taken prior to a UV regularization, the quasi-observable $\tilde{\mathcal{O}}$ becomes the parton observable $\mathcal{O}$ by construction. However, what one can calculate in practice is the quasi-observable $\tilde{\mathcal{O}}$ at large but finite P^z with UV regularization imposed first. This is the case in lattice computations. The difference between $\mathcal{O}$ and $\tilde{\mathcal{O}}$ is just the order of limits. This is similar to an effective field theory set-up. The difference is that here the role of perturbative degrees of freedom is played by the large momentum of the external state, hence it cannot be arranged into a Lagrangian formalism. Nevertheless, one can bridge the quasi- and parton observables through

$$\tilde{\mathcal{O}}(P^z/\Lambda) = Z\left(P^z/\Lambda, \mu/\Lambda\right)\mathcal{O}(\mu) + \frac{c_2}{(P^z)^2} + \frac{c_4}{(P^z)^4} + \cdots, \tag{10}$$

where Λ is a UV cutoff imposed on the quasi-observable, and c_i's are higher-twist contributions suppressed by powers of P^z. That is, the quasi-observable $\tilde{\mathcal{O}}(P^z/\Lambda)$ can be factorized into the parton observable $\mathcal{O}(\mu)$ and a matching coefficient Z, up to power suppressed corrections. Taking the $P^z \to \infty$ limit does not change the infrared (IR) behavior of the observable, but only affects its UV behavior. Therefore $\mathcal{O}(\mu)$ captures all the IR physics in $\tilde{\mathcal{O}}(P^z/\Lambda)$, and the matching coefficient Z is completely perturbative.

An explicit example of Eq. (10) is presented in Refs. 40, 41 for the case of parton distribution functions (PDFs), where the factorization formula has a convolution form, and the matching coefficients were calculated at the leading logarithmic order. Using these results, the first direct lattice calculation of the isovector sea-quark parton distributions has been available recently.[42] A similar factorization formula was also proposed in Ref. 43 to extract PDFs from lattice QCD calculations based on QCD factorization of lattice "cross sections".

Within the LaMET framework, we can start with suitable quasi-observables to calculate the proton spin content. According to our discussions above, these quasi-observables can be defined as the free-field QCD AMOs in a "physical" gauge condition that has the correct WW approximation in the IMF limit.[24] A possible choice of the "physical" gauge condition is the expression in terms of nonlocal operators introduced by Chen *et al.*:[13,44]

$$\vec{J}_{\mathrm{QCD}} = \int d^3x\ \psi^\dagger \frac{\vec{\Sigma}}{2}\psi + \int d^3x\ \psi^\dagger \vec{x} \times (-i\vec{\nabla} - g\vec{A}_\parallel)\psi$$
$$+ \int d^3x\ \vec{E}_a \times \vec{A}_\perp^a + \int d^3x\ E_a^i\ (\vec{x} \times \vec{\nabla})A_\perp^{i,a}, \tag{11}$$

where $\vec{x}$ are the spatial coordinates, and $\vec{A}$ is decomposed into a pure-gauge part $\vec{A}_\parallel$ and a physical part $\vec{A}_\perp$ which satisfy (see also Ref. 45)

$$\partial^i A_\parallel^{j,a} - \partial^j A_\parallel^{i,a} - g f^{abc} A_\parallel^{i,b} A_\parallel^{j,c} = 0 \ ,$$

$$\partial^i A_\perp^i - ig[A^i, A_\perp^i] = 0, \tag{12}$$

so that each term in Eq. (11) is gauge invariant. From Eq. (12), one can show that in the Coulomb gauge $\vec{\nabla} \cdot \vec{A} = 0$, $\vec{A}_\perp$ equals $\vec{A}$ order by order in perturbation theory. Therefore, Eq. (11) corresponds to choosing the Coulomb gauge as the "physical" gauge.

It has been shown in Ref. 23 that $\vec{E}_a \times \vec{A}_\perp^a$ in Eq. (11) is equivalent to the total gluon spin operator in the IMF limit. It is easy to see that the other nonlocal terms in Eq. (11) also have the correct WW approximation as the parton OAM. Therefore, we can choose the nonlocal operators in Eq. (11) as the quasi-observables for parton angular momentum in the LaMET approach.

The advantage of the expression in Eq. (11) is that it is time independent and thus allows for a direct calculation in lattice QCD. Suppose we evaluate the matrix elements of these quasi-observables with finite momentum P^z, we should have

$$\frac{1}{2} = \frac{1}{2}\Delta\widetilde{\Sigma}(\mu, P^z) + \Delta\widetilde{G}(\mu, P^z) + \Delta\widetilde{L}_q(\mu, P^z) + \Delta\widetilde{L}_g(\mu, P^z), \tag{13}$$

where the P^z-dependence is expected since Eq. (11) is a frame-dependent expression.[23] Following the effective theory argument above, we can relate these quasi-observables to the corresponding parton observables through the following factorization formula:

$$\Delta\widetilde{\Sigma}(\mu, P^z) = \Delta\Sigma(\mu),$$

$$\Delta\widetilde{G}(\mu, P^z) = z_{qg}\Delta\Sigma(\mu) + z_{gg}\Delta G(\mu) + O\left(\frac{M^2}{(P^z)^2}\right),$$

$$\Delta\widetilde{L}_q(\mu, P^z) = P_{qq}\Delta L_q(\mu) + P_{gq}\Delta L_g(\mu)$$

$$+ p_{qq}\Delta\Sigma(\mu) + p_{gq}\Delta G(\mu) + O\left(\frac{M^2}{(P^z)^2}\right),$$

$$\Delta\widetilde{L}_g(\mu, P^z) = P_{qg}\Delta L_q(\mu) + P_{gg}\Delta L_g(\mu)$$

$$+ p_{qg}\Delta\Sigma(\mu) + p_{gg}\Delta G(\mu) + O\left(\frac{M^2}{(P^z)^2}\right), \tag{14}$$

where M is the proton mass, and all the matrix elements are renormalized in the $\overline{\text{MS}}$ scheme. $\Delta\widetilde{\Sigma}(\mu, P^z)$ is the same as $\Delta\Sigma(\mu)$ because the quark spin operator is frame independent and has the same matrix elements in the Coulomb and light-cone gauges. The z_{ij}, P_{ij} and p_{ij}'s are the matching coefficients to be calculated in perturbative QCD.

All the matching coefficients in Eq. (14) at one-loop order has been calculated in Ref. 46:

$$P_{qq} = 1 + \frac{\alpha_S C_F}{4\pi}\left(-2\ln\frac{(P^z)^2}{\mu^2} + R_3\right), \quad P_{gq} = 0,$$

$$P_{qg} = \frac{\alpha_S C_F}{4\pi}\left(2\ln\frac{(P^z)^2}{\mu^2} - R_3\right), \qquad P_{gg} = 1,$$

$$p_{qq} = \frac{\alpha_S C_F}{4\pi}\left(-\frac{1}{3}\ln\frac{(P^z)^2}{\mu^2} + R_4\right), \qquad p_{gq} = 0, \tag{15}$$

$$p_{qg} = \frac{\alpha_S C_F}{4\pi}\left(-\ln\frac{(P^z)^2}{\mu^2} - R_1 - R_4\right),$$

$$p_{gg} = \frac{\alpha_S C_A}{4\pi}\left(-\frac{7}{3}\ln\frac{(P^z)^2}{\mu^2} - R_2\right),$$

where

$$C_F = (N_c^2 - 1)/2N_c, \; C_A = N_c,$$

$$R_1 = \tfrac{8}{3}\ln 2 - \tfrac{64}{9}, \quad R_2 = \frac{14}{3}\ln 2 - \frac{121}{9},$$

$$R_3 = -4\ln 2 + \tfrac{28}{3}, \quad R_4 = -\frac{2}{3}\ln 2 + \frac{13}{9}, \tag{16}$$

with N_c being the number of colors.

With the above results, we are able to convert the quasi-observables in Eq. (11) evaluated at a large finite momentum to the parton spin and OAM in IMF. This can be done by a simple inversion of Eq. (14). Although for realistic lattice computations the above matching coefficients have to be recalculated using lattice perturbation theory,[47] the leading logarithmic term of the matching coefficients is independent of the regularization scheme, and therefore is the same in dimensional and lattice regularizations. Our one-loop matching coefficients can thus be used for an approximate lattice computation of parton angular momentum to leading logarithmic accuracy.

In summary, we have justified the physical significance of the Jaffe-Manohar spin sum rule as a result of the WW approximation in the IMF. The key distinction between the Jaffe-Manohar and Ji's sum rules is that the former is defined in the IMF while the latter is frame independent. The quark and gluon OAM in the Jaffe-Manohar sum rule can be measured through twist-three observables from two-photon processes such as DVCS. In addition, we have shown how to obtain the partonic contributions to proton spin using the LaMET approach. The solution is a perturbative factorization formula that allows us to extract the parton spin and OAM in the IMF from lattice QCD calculations with a finite but large proton momentum. With such developments, we can eventually compare the parton spin and OAM in theory and experiment.

We thank J. -W. Chen for useful discussions about the gauge-invariant gluon spin operator. We also thank M. Glatzmaier and K. -F. Liu for discussions on matching in lattice perturbation theory. This work was partially supported by the U.S. Department of Energy Office of Science, Office of Nuclear Physics under Award Number DE-FG02-93ER-40762 and a grant (No. 11DZ2260700) from the Office of Science and Technology in Shanghai Municipal Government, and grants by the National Science Foundation of China (No. 11175114, No. 11405104).

References

1. J. Ashman *et al.* [European Muon Collaboration], Phys. Lett. B **206**, 364 (1988).
2. J. Ashman *et al.* [European Muon Collaboration], Nucl. Phys. B **328**, 1 (1989).
3. C. A. Aidala, S. D. Bass, D. Hasch and G. K. Mallot, Rev. Mod. Phys. **85**, 655 (2013) [arXiv:1209.2803 [hep-ph]].
4. D. de Florian, R. Sassot, M. Stratmann and W. Vogelsang, Phys. Rev. D **80**, 034030 (2009) [arXiv:0904.3821 [hep-ph]].
5. D. de Florian, R. Sassot, M. Stratmann and W. Vogelsang, Phys. Rev. Lett. **113**, no. 1, 012001 (2014) [arXiv:1404.4293 [hep-ph]].
6. J. Dudek, R. Ent, R. Essig, K. S. Kumar, C. Meyer, R. D. McKeown, Z. E. Meziani and G. A. Miller *et al.*, Eur. Phys. J. A **48**, 187 (2012) [arXiv:1208.1244 [hep-ex]].
7. A. Accardi, J. L. Albacete, M. Anselmino, N. Armesto, E. C. Aschenauer, A. Bacchetta, D. Boer and W. Brooks *et al.*, arXiv:1212.1701 [nucl-ex].
8. R. L. Jaffe and A. V. Manohar, *Nucl. Phys. B* **337**, 509 (1990).
9. X. -D. Ji, *Phys. Rev. Lett.* **78**, 610 (1997) [hep-ph/9603249].
10. X. D. Ji, *Phys. Rev. D* **55**, 7114 (1997) [hep-ph/9609381].
11. P. Hoodbhoy, X. D. Ji and W. Lu, *Phys. Rev. D* **59**, 014013 (1998) [hep-ph/9804337].
12. X. Ji, X. Xiong and F. Yuan, *Phys. Lett. B* **717**, 214 (2012) [arXiv:1209.3246 [hep-ph]].
13. X. S. Chen, X. F. Lu, W. M. Sun, F. Wang and T. Goldman, Phys. Rev. Lett. **100**, 232002 (2008) [arXiv:0806.3166 [hep-ph]].
14. M. Wakamatsu, Phys. Rev. D **83**, 014012 (2011) [arXiv:1007.5355 [hep-ph]].
15. Y. Hatta, Phys. Rev. D **84**, 041701 (2011) [arXiv:1101.5989 [hep-ph]].
16. B. H. Zhou and Y. C. Huang, Phys. Rev. D **84**, 047701 (2011).
17. E. Leader and C. Lorc, Phys. Rept. **541**, 163 (2014) [arXiv:1309.4235 [hep-ph]].
18. R. L. Jaffe, Phys. Lett. B **365**, 359 (1996) [hep-ph/9509279].
19. X. -S. Chen and F. Wang, hep-ph/9802346;
    F. Wang, W. -M. Sun and X. -F. Lu, hep-ph/0510285.
20. E. Leader, Phys. Rev. D **83**, 096012 (2011) [Erratum-ibid. D **85**, 039905 (2012)] [arXiv:1101.5956 [hep-ph]].
21. P. Hoodbhoy and X. D. Ji, Phys. Rev. D **60**, 114042 (1999) [hep-ph/9908275].
22. X. Ji, Y. Xu, and Y. Zhao, JHEP **1208**, 082 (2012).
23. X. Ji, J. H. Zhang and Y. Zhao, Phys. Rev. Lett. **111**, 112002 (2013) [arXiv:1304.6708 [hep-ph]].
24. Y. Hatta, X. Ji and Y. Zhao, Phys. Rev. D **89**, 085030 (2014) [arXiv:1310.4263 [hep-ph]].
25. V. B. Berestetskii, L. P. Pitaevskii, and E.M. Lifshitz, "Quantum Electrodynamics," Second Edition: Volume 4 in Courses in Theoretical Physics, Butterworth-Heineman, 1982.
26. X. -D. Ji, J. Tang and P. Hoodbhoy, Phys. Rev. Lett. **76**, 740 (1996) [hep-ph/9510304].
27. J. D. Jackson, *Classical Electrodynamics* (John Wiley & Sons, New York, 1999), 3rd ed..
28. Y. Hatta, Phys. Lett. B **708**, 186 (2012) [arXiv:1111.3547 [hep-ph]].

29. X. Ji, X. Xiong and F. Yuan, Phys. Rev. D **88**, no. 1, 014041 (2013) [arXiv:1207.5221 [hep-ph]].
30. X. Ji, X. Xiong and F. Yuan, Phys. Rev. Lett. **109**, 152005 (2012) [arXiv:1202.2843 [hep-ph]].
31. I. V. Anikin, B. Pire and O. V. Teryaev, Phys. Rev. D **62**, 071501 (2000) [hep-ph/0003203].
32. A. V. Belitsky and D. Mueller, Nucl. Phys. B **589**, 611 (2000) [hep-ph/0007031].
33. P. Hagler *et al.* [LHPC and SESAM Collaborations], Phys. Rev. D **68**, 034505 (2003) [hep-lat/0304018].
34. J. D. Bratt *et al.* [LHPC Collaboration], Phys. Rev. D **82**, 094502 (2010) [arXiv:1001.3620 [hep-lat]].
35. D. Brommel *et al.* [QCDSF-UKQCD Collaboration], PoS LAT **2007**, 158 (2007) [arXiv:0710.1534 [hep-lat]].
36. C. Alexandrou, M. Constantinou, S. Dinter, V. Drach, K. Jansen, C. Kallidonis and G. Koutsou, Phys. Rev. D **88**, no. 1, 014509 (2013) [arXiv:1303.5979 [hep-lat]].
37. C. Alexandrou, J. Carbonell, M. Constantinou, P. A. Harraud, P. Guichon, K. Jansen, C. Kallidonis and T. Korzec *et al.*, Phys. Rev. D **83**, 114513 (2011) [arXiv:1104.1600 [hep-lat]].
38. M. Deka, T. Doi, Y. B. Yang, B. Chakraborty, S. J. Dong, T. Draper, M. Glatzmaier and M. Gong *et al.*, Phys. Rev. D **91**, no. 1, 014505 (2015) [arXiv:1312.4816 [hep-lat]].
39. X. Ji, Sci. China Phys. Mech. Astron. **57**, no. 7, 1407 (2014) [arXiv:1404.6680 [hep-ph]].
40. X. Ji, Phys. Rev. Lett. **110**, 262002 (2013) [arXiv:1305.1539 [hep-ph]].
41. X. Xiong, X. Ji, J. H. Zhang and Y. Zhao, Phys. Rev. D **90**, 014051 (2014) [arXiv:1310.7471 [hep-ph]].
42. H. -W. Lin, J. -W. Chen, S. D. Cohen and X. Ji, arXiv:1402.1462 [hep-ph].
43. Y. Q. Ma and J. W. Qiu, arXiv:1404.6860 [hep-ph].
44. X. -S. Chen, W. -M. Sun, X. -F. Lü, F. Wang, and T. Goldman, Phys. Rev. Lett. **103**, 062001 (2009).
45. R. P. Treat, J. Math. Phys. **13**, 1704 (1972) [Erratum-ibid. **14**, 1296 (1973)].
46. X. Ji, J. H. Zhang and Y. Zhao, Phys. Lett. B **743**, 180 (2015) [arXiv:1409.6329 [hep-ph]].
47. S. Capitani, Phys. Rept. **382**, 113 (2003) [hep-lat/0211036].

Spin Physics (SPIN2014)
International Journal of Modern Physics: Conference Series
Vol. 40 (2016) 1660002 (10 pages)
© The Author(s)
DOI: 10.1142/S2010194516600028

World Scientific
www.worldscientific.com

Status of Theory and Experiment in Hadronic Parity Violation

W. M. Snow

Department of Physics, Indiana University
Bloomington, IN 47401 USA
wsnow@indiana.edu

M. W. Ahmed

Department of Physics, North Carolina Central University
Durham, NC 27710 USA
ahmed@tunl.duke.edu

J. D. Bowman

Physics Division, Oak Ridge National Lab
Oak Ridge, TN 37831 USA
bowmanjd@ornl.gov

C. Crawford

Department of Physics, University of Kentucky
Lexington, KY 40506 USA
crawford@pa.uky.edu

N. Fomin

Department of Physics, University of Tennessee
Knoxville, TN 37916 USA
nfomin@utk.edu

H. Gao

Department of Physics, Duke University
Durham, NC 27710 USA
gao@phy.duke.edu

M. T. Gericke

Department of Physics, University of Manitoba
Winnipeg, MB R3T 2N2 Canada
Michael.Gericke@umanitoba.ca

V. Gudkov

Department of Physics, University of South Carolina
Columbia, SC 29208 USA
gudkov@sc.edu

B. R. Holstein

Department of Physics, University of Massachusetts
Amherst, MA 01003 USA
Holstein@physics.umass.edu

C. R. Howell

Department of Physics, Duke University
Durham, NC 27710 USA
howell@tunl.duke.edu

P. Huffman

Department of Physics, North Carolina State University
Raleigh, NC 27607 USA
paulhuffman@ncsu.edu

W. T. H. van Oers

TRIUMF and Department of Physics, University of Manitoba
Winnipeg, MB R3T 2N2 Canada
vanoers@triumf.ca

S. Penttilä

Physics Division, Oak Ridge National Lab
Oak Ridge, TN 37831 USA
penttilasi@ornl.gov

Y. K. Wu

Department of Physics, Duke University
Durham, NC 27710 USA
wu@fel.duke.edu

Published 29 February 2016

Hadronic parity violation uses quark-quark weak interactions to probe nonperturbative strong interaction dynamics through two nonperturbative QCD scales: Λ_{QCD} and the fine-tuned MeV scales of NN bound states in low energy nuclear physics. The current and projected availability of high-intensity neutron and photon sources coupled with ongoing experiments and continuing developments in theoretical methods provide the opportunity to greatly expand our understanding of hadronic parity violation in few-nucleon systems. The current status of these efforts and future plans are discussed.

Keywords: Parity violation; QCD; lattice gauge theory; effective field theory.

PACS numbers: 21.30.Cb, 11.30.Er, 24.70.+s, 13.75.Cs

1. Introduction

Hadronic parity violation offers a window into one of the least understood sectors of the Standard Model, the neutral-current nonleptonic weak interactions. Weak interactions are well-understood at the quark level, and neutral-current contributions are highly suppressed in flavor-changing hadronic decays due to the absence of tree level flavor changing neutral currents in the Standard Model. The only experimentally accessible approach is thus to study the parity-violating (PV) component of nucleon-nucleon (NN) interactions and its manifestation in nuclear systems.

Hadronic parity-violation is also a unique probe of nonperturbative dynamics of the strong interactions at low energies. With the weak interactions known at the quark level, their manifestation in PV NN interactions is the result of an interference with the nonperturbative strong effects that confine the quarks in the nucleons and lead to the residual strong NN interaction. The large mass of the weak gauge bosons implies that the range of the weak interactions ($1/M_{W,Z} \approx 0.002$ fm) is very small compared to the size of the nucleon (roughly 1 fm). The weak NN interactions are therefore sensitive to short-distance correlations between quarks inside the nucleon, and their manifestations at the hadronic level provide information about these correlations without the need of an external probe.

For low-energy nuclear reactions, hadronic parity violation can be described in terms of interactions between nucleons. The complex interplay between weak and nonperturbative strong physics is encoded in five PV NN amplitudes. Experimental observables are linear combinations of the NN amplitudes, and the relative weightings of these NN amplitudes must be calculated to map out the PV landscape with an appropriate suite of measurements. This information should be derived from two- and few-nucleon systems to avoid the still-unknown complexities of nuclear structure.

Not enough experimental information from two- and few-nucleon systems is available yet to reliably determine all of the weak NN amplitudes. However, the existing and projected availability of high-intensity sources of neutrons and photons combined with ongoing technical improvements in the control of systematic uncertainties make new experiments possible. In parallel, the theoretical tools needed for a reliable and systematic analysis and interpretation of the experimental results continue to be developed. In particular, effective field theory (EFT) methods that are applicable to few-nucleon systems provide model-independent results which by construction obey the known symmetries of QCD along with theoretically justifiable error estimates.

Ultimately, the weak NN amplitudes have to be related to Standard Model parameters by calculations at the quark level. They provide a qualitatively distinct piece of low-energy data that any nonperturbative QCD calculation has to reproduce. The relative size of the various amplitudes can provide a hint at the underlying dynamics. The first lattice QCD calculation of a hadronic parity violation amplitude appeared recently.[1] The importance of lattice QCD calculations in determining PV nucleon interactions was also highlighted in the 2009 workshop "Forefront Questions in Nuclear Science and the Role of Computing at the Extreme Scale" .[2] Since little is known about the PV nucleon couplings, they provide an opportunity to utilize lattice QCD and EFT to make predictions in the nonperturbative regime, which in turn can be experimentally tested.

2. Current Experimental Status

After a decade in which there was almost no new experimental information on NN weak interaction amplitudes, we now anticipate three experimental results on NN parity violation in two and few-nucleon systems which promise to greatly improve our knowledge of this sector. A 2011 NSAC review of the subfield of fundamental neutron physics identified NPDGamma as the highest priority experiment in NN weak interaction physics. NPDGamma was selected as the first experiment to run on the new FnPB facility for neutron physics at the Spallation Neutron Source at Oak Ridge National Lab. NPDGamma measures the parity-odd asymmetry of the 2.2 MeV photons from polarized slow neutron capture on protons and is sensitive to a S-P NN weak transition amplitude in the $\Delta I = 1$ channel corresponding to weak pion exchange. NPDGamma finished taking data in mid-2014 and data analysis is in progress. The preliminary result for the parity-violating asymmetry A_γ is that it is small with a statistical error of ~ 13 ppb and with negligible systematic error.[3] This result is consistent with both the longstanding result from experiments and theoretical analysis of parity violation in ^{18}F, which sees no evidence for a $\Delta I = 1$ NN weak amplitude from pion exchange, with a recent pioneering lattice calculation,[1] and with $1/N_c$ arguments.[4,5]

This result will have a number of implications. For the first time, the size of the S-P NN weak transition amplitude in the $\Delta I = 1$ channel corresponding to weak pion exchange will soon be tightly constrained with data from a two-nucleon system. The size of this amplitude is small compared to the size of the $\Delta I = 0$ NN weak amplitudes which have been constrained by previous experimental observations of parity violation in proton scattering. The four-quark operators responsible for NN weak interactions in $\Delta I = 0$ and $\Delta I = 1$ channels have similar strength at the electroweak scale, and the perturbative QCD evolution of these operators from the electroweak scale down to Λ_{QCD} does not change this. We therefore suspect that this large difference is a sign of interesting nonperturbative QCD dynamics. We also expect that the $\Delta I = 1$ NN weak amplitudes are dominated by quark-quark neutral currents as the charged current contributions at the electroweak scale are suppressed by a factor of V_{us}^2. If so, the NPDGamma result implies that quark-quark neutral currents seem to be suppressed in NN weak transition amplitudes.

The long range of the $\Delta I = 1$ NN weak amplitude corresponding to weak pion exchange contributes to almost all parity-odd observables in two- and few-nucleon systems. The NPDGamma result that this amplitude is both small and tightly constrained will make it much easier to extract new information on the other NN weak transition amplitudes from additional experiments in two- and few-nucleon systems.

Important progress has been made on two other parity violation experiments in few-nucleon systems. An experiment to measure the proton asymmetry in polarized slow neutron capture on ^{3}He (the $\vec{s_n} \cdot \vec{k_p}$ correlation in $\vec{n} + {}^3\text{He} \rightarrow {}^3\text{H} + p$) is taking data on the FnPB at the SNS with a projected statistical accuracy of 1.6×10^{-8}.[6,7] The experiment has measured a small (ppm) parity-conserving asymmetry $(\vec{s_n} \cdot (\vec{k_n} \times \vec{k_p}))$ correlation. The predicted PV asymmetry is 1.2×10^{-7}. A statistically-limited null result in polarized slow neutron spin rotation in $\vec{n} + {}^4\text{He}$ was measured on the NG-6 beamline at NIST,[8] and a new spin rotation apparatus is under construction which is proposed to run on the new more intense NG-C beam at NIST. The projected precision of both of these experiments is enough to tightly constrain linear combinations of parity-odd transition amplitudes. Also, in both of these parity-odd observables the relative signs of the $\Delta I = 0$ and $\Delta I = 1$ transition amplitudes are different from the isospin-conjugate $\vec{p} + {}^4\text{He}$ system, where a nonzero parity-odd asymmetry was measured earlier at PSI. Data on parity violation in proton-proton scattering already exists as well. We therefore anticipate that the addition of the new results from these three experiments should greatly reduce the uncertainly on the values of these parity-odd amplitudes.

3. Future Experimental Opportunities

After these three experiments are completed over the next few years, we will have a much better feeling for what should be the highest priority for future experiments. The NPDGamma experiment is not limited by systematic errors, and it is anticipated that the same might well be true for the n-^{3}He and n+^{4}He measurements. It is therefore likely that it would be interesting to repeat one or more of these experiments should a higher intensity slow neutron beamline become available. A very interesting slow neutron beamline for experiments of relevance to nuclear/particle/astrophysics/cosmology has recently been proposed for the European Spallation Source (ESS) now under construction in Lund, Sweden. The ESS is designed as a so-called "long-pulse" spallation neutron source with a design power of 5 MW, larger than the SNS by a factor of 5. Combined with some improvements in the brightness in slow neutron moderators, it promises to deliver more than an order of magnitude more polarized slow neutrons than any other pulsed neutron source. The pulsed nature of the source is very important for these experiments to successfully isolate systematic errors as one gets neutron energy information from time-of-flight and many possible systematic effects depend on the neutron kinetic energy. If this beamline is realized it will be possible to either improve the statistical accuracy of one or more of the experiments mentioned above or perhaps to measure NN parity violation in a different system such as polarized slow neutron capture on deuterium. By the time that this facility becomes available we will be able to make a much more intelligent choice of what experiment to pursue given the results of the ongoing experiments combined with the ongoing progress from theory.

There is another interesting scientific opportunity in experimental NN weak interaction physics. This is the measurement of the parity-odd helicity dependence of the photodisintegration of the deuteron near threshold in the $\vec{\gamma} + D \to n + p$ reaction. Together with the parity-odd longitudinal asymmetry in $\vec{p} + p$ scattering, this observable is sensitive to the experimentally elusive $\Delta I = 2$ NN weak amplitude. This amplitude is especially interesting from a theoretical point of view as it comes from one $\Delta I = 2$ effective 4-quark operator above Λ_{QCD} and it is the most accessible channel for a calculation from the Standard Model using lattice gauge theory. This experiment could be done in the future at a high-intensity polarized gamma source. The proposed upgrades to the HIγS facility at Duke in beam intensity and in the capability for rapid and controlled helicity flipping of the photon beam could make it possible to conduct this experiment. Such an upgrade could also enable other parity violation experiments in few nucleon systems.

4. Theoretical Status

Traditionally PV NN interactions have been described in a meson-exchange picture in analogy to many parity-conserving (PC) NN models. Most widely used has been the DDH formulation,[9] which also provides estimated ranges for the PV couplings appearing in the model. The available experimental results from two- to many-nucleon systems can be described in terms of four DDH parameters, provided a one-body PV potential is derived in the case of heavy nuclei.[10] Only one $\Delta I = 1$ parameter is needed to describe the experimental data.

Over the last decades, EFTs have emerged as alternatives to phenomenological models of the NN interactions, both in the PC and PV sectors. EFTs are model independent, and provide systematic theoretical error estimates and a connection to QCD via symmetry principles. The use of EFTs to study hadronic parity violation goes back to the 90s,[11] with a comprehensive analysis of PV Lagrangians given in Ref. 12. EFTs have now been applied to PV observables in two-, three-, and four-nucleon systems (see Ref. 13 for a recent review), including the corresponding observables discussed in the previous sections. Such calculations play an important role in assessing whether measurements are feasible and, if so, in the planning and the subsequent analysis of the corresponding experiments. EFT methods continue to be developed, with a particular emphasis on the application to few-nucleon systems. $p + {}^4$He scattering at low energy was recently calculated using chiral EFT potentials.[14] PV three-nucleon interactions were shown to be 10% effects.[15] This means that at very low energy the five PV low-energy constants (LECs) are sufficient to describe parity violation in few-nucleon systems. The PV asymmetry in $\vec{n} + {}^3$He $\to {}^3$H $+ p$ was calculated employing chiral EFT for both PC and PV interactions. This calculation is essential for the interpretation of the upcoming measurement of this observable at the FnPB. The relationships and connections between two different descriptions of NN weak interactions, the model-dependent DDH approach and the various versions of EFT have been greatly clarified onver the last few years.

There has been tremendous progress in the development of lattice QCD as a reliable tool to compute basic properties of low-energy nuclear and hadronic physics. Lattice QCD is now being utilized to make theoretical predictions of important low-energy strong interaction amplitudes which are either challenging or currently not possible to measure experimentally. The pioneering calculation of the PV pion-nucleon coupling in Ref. 1 showed that lattice QCD can have significant impact on our understanding of hadronic parity violation. The resulting value of $h_\pi^{1,\text{con}} = (1.099 \pm 0.505^{+0.058}_{-0.064}) \times 10^{-7}$ is in agreement with present experimental bounds. It is possible to utilize lattice QCD to make predictions of the yet undetermined PV coupling constants appearing in the PV EFT, which in turn can be tested experimentally. This program could establish a quantitative connection through two nonperturbative strong interaction scales (Λ_{QCD} and the $\sim$ MeV scale of NN bound state dynamics) from quark-level interactions to observables in few-nucleon systems. The demonstration of success in this very nontrivial goal would represent an impressive milestone in our understanding of the strongly interacting limit of QCD.

5. Theoretical Opportunities

There are a number of opportunities for theoretical progress in hadronic parity violation. The ongoing improvements in the application of EFT methods to few-nucleon systems put consistent calculations of the PV asymmetry in $\vec{p} + {}^4\text{He}$ (which has been measured) and the spin rotation angle in $\vec{n} + {}^4\text{He}$ (which is planned for NIST) within reach. The power counting analysis of EFTs is based on the assumption that the couplings are natural in size. Experimental results that reveal a pattern of values for the LECs different from O(1) are an indication of nontrivial QCD dynamics. An updated analysis of the EFT power counting may be required in that case. Lattice QCD plays an equally important role in establishing the connection between the fundamental Standard Model parameters to weak NN amplitudes. The $\Delta I = 2$ amplitude presents an especially interesting target for investigation because it only receives contributions from connected quark diagrams (not the computationally demanding disconnected loop diagrams) and the isotensor nature of the operator prevents mixing under renormalization. First lattice QCD results on this amplitude from the CalLat Collaboration are anticipated in the near future. The determination of this quantity from experiment is difficult and currently not well constrained. Our ignorance of the size of this $\Delta I = 2$ amplitude also contributes significantly to the uncertainty in the determination of the $\Delta I = 0$ amplitude from p-p scattrering.[16] This amplitude presents a very exciting opportunity to use lattice QCD and EFT to make a precise prediction of a NN weak amplitude directly from the Standard Model.

6. HIγS2: The Compton γ-Ray Source

The measurement of the parity-violating asymmetry of photodisintegration of the deuteron to an accuracy of 10^{-8} requires a high intensity circularly polarized γ-ray beam with high polarization ($> 90\%$) and narrow beam energy spread (< 100 keV FWHM at about 2.4 MeV). The current facility best suited to deliver the γ-ray beam for this experiment is the High Intensity Gamma-ray Source (HIγS) at the Triangle Universities Nuclear Laboratory (TUNL). The γ-ray beam at HIγS is produced by Compton backscattering of the photons inside the optical cavity of a storage-ring based Free-Electron Laser from circulating electron bunches. The current capabilities of HIγS are: (1) delivery of γ-ray beams on target in the energy of 1 to 100 MeV with beam energy spread selected by collimation as low as about 1% FWHM, (2) a beam intensity on target with 5% energy spread as high as 10^9 gammas/s, (3) linear or circular beam polarization with magnitude of polarization greater than 95%. To carry out the proposed parity violation experiment will require an upgrade of HIγS to increase the beam intensity by about two orders of magnitude and to provide the capability of fast reversal of the beam helicity with precision control and diagnostics of the spatial and energy distributions of the beam. A two-stage upgrade of HIγS is envisaged. The first stage is an upgrade of the electron injector system to increase the average charge injection rate into the storage ring and to enhance the overall reliability of the accelerator drivers. This stage is motivated by the Compton-scattering research program at HIγS and is also needed for the parity-violation experiments. This part of the accelerator upgrade will start around 2018. The second stage in the upgrade is the installation of an optical cavity in the straight section of the storage ring that is driven by an external laser. The photons in the optical cavity will produce γ-rays by Compton scattering from the electron bunches circulating in the storage ring. The schedule for starting construction is around 2021.

A schematic diagram of HIγS2 is shown in Fig. 1. The HIγS2 γ-ray beam is produced by colliding a high-current electron beam with photons inside an optical cavity that is driven by a powerful external laser. The electron beam is provided by the storage ring which will be operated with 32 electron bunches with a total current up to about 500 mA. An average optical power of about 10 to 20 KW will be built up inside the high-finesse Fabry-Perot resonator. Compared with the FEL driven HIγS, a higher average intracavity laser power and significantly reduced beam sizes at the collision point make it possible for the HIγS2 to achieve a total gamma flux of 10^{11} to 10^{12} γ/s in the energy range of 2 to 12 MeV. In addition to the substantial intensity increase, the HIγS2 facility will produce γ-ray beams with a better monochromaticity and the ability to rapidly switch the γ-ray beam helicity at a rate of tens of Hz or higher. The fast helicity switch of the γ-ray beam will be realized by changing the polarization of the laser beam outside the Fabry-Perot resonator using polarizing optics such as Pockels cells and wave plates.

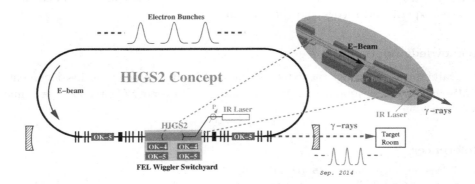

Fig. 1. The schematic layout of the HIγS2 Compton γ-ray source. The HIγS2 is located in one of the two long straight sections of the electron storage ring. Guided by a set of deflection magnets, the electron beam enters and exits the high-power Fabry-Perot resonator to collider with the photon pulse in the center of the resonator. The polarization of the γ-ray beam is rapidly switchable by changing the polarization of the laser beam from the external infrared drive laser using polarizing optics.

7. Summary and Conclusions

Quark-quark weak interactions in the Standard Model induce parity-odd nucleon-nucleon interactions. Hadronic parity violation provides a unique probe of neutral-current nonleptonic weak interactions as well as of nonperturbative strong dynamics. The preliminary result from the NPDGamma experiment at ORNL, the first new sensitive experimental result in the NN system in many years, shows that at least one NN weak amplitude is smaller than expected on symmetry arguments alone. It strongly supports the longstanding suspicion from analysis of previous data that NN weak amplitudes are sensitive to poorly-understood aspects of nonperturbative QCD dynamics. This observation is intellectually exciting as it shows that NN weak interaction measurements could open a unique window on nonperturbative QCD physics. The availability of high-intensity sources of neutrons and photons such as the FnPB at the Spallation Neutron Source at ORNL, the new NG-C neutron beam at the NCNR at NIST, the European Spallation Source under construction in Sweden, and an upgraded HIγS gamma facility at TUNL along with the controls for systematic uncertainties possible at these facilities provides new opportunities to greatly improve our experimental information on hadronic parity violation. Effective field theory methods are ideally suited to analyze and interpret the experimental results in this field. The recent progress in the application of effective field theories to few-nucleon systems should make it possible to determine PV observables in processes involving as many as five nucleons. The weak amplitudes at the nucleon level should ultimately be connected to the parameters of the Standard Model. An understanding of the neutral-current weak interaction and its manifestation in hadronic parity violation from the Standard Model will require a coordinated effort between experiment, theory and computational physics.

Acknowledgments

We gratefully acknowledge the extensive input of M. R. Schindler on this document. WMS acknowledges support through the Indiana University Center for Spacetime Symmetries.

References

1. J. Wasem, *Phys. Rev. C* **85**, 022501 (2012).
2. Scientific Grand Challenges: Forefront Questions in Nuclear Science and the Role of Computing at the Extreme Scale, in *Workshop sponsored by the DOE Office of Nuclear Physics and the Office of Advanced Scientific Computing*, 2009.
3. J. D. Bowman, presentation at the Electron-Nucleus Scattering XIII conference, Elba, Italy. June 2014.
4. N. Kaiser and U. G. Meissner, *Nucl. Phys. A* **499**, 699 (1989).
5. D. R. Phillips, D. Samart, and C. Schat, *Phys. Rev. Lett.* **99**, (2014).
6. V. Gudkov, *Phys. Rev. C* **82**, 065502 (2010).
7. M. Vivani *et al*, *Phys. Rev. C* **89**, 064004 (2014).
8. W. M. Snow *et al.*, *Phys.Rev. C* **83**, 022501 (2011).
9. B. Desplanques, J. F. Donoghue, and B. R. Holstein, *Annals Phys. (NY)* **124**, 449 (1980).
10. J. D. Bowman, presentation at INT program on "Electric Dipole Moments and CP Violations", (2007).
11. D. B. Kaplan and M. J. Savage, *Nucl. Phys. A* **556**, 653 (1993).
12. S. L. Zhu, C. M. Maekawa, B. R. Holstein, and M. J. Ramsey-Musolf, *Nucl. Phys. A* **748**, 435 (2005).
13. M. R. Schindler and R. P. Springer, *Prog. Part. Nucl. Phys.* **72**, 1 (2013).
14. G. Hupin, S. Quaglioni, and P. Navratil, *Phys. Rev. C* **90**, 061601 (2014).
15. H. W. Griesshammer and M. R. Schindler, *Eur. Phys. J A* **46**, 73 (2010).
16. W. C. Haxton and B. Holstein, *Prog. Part. Nucl. Phys.* **71**, 185 (2013).

Spin Physics (SPIN2014)
International Journal of Modern Physics: Conference Series
Vol. 40 (2016) 1660003 (10 pages)
© The Author(s)
DOI: 10.1142/S201019451660003X

Beam Polarization at the ILC: Physics Case and Realization

Annika Vauth

Deutsches Elektronen-Synchrotron DESY
Notkestraße 85, 22607 Hamburg, Germany
annika.vauth@desy.de

Jenny List

Deutsches Elektronen-Synchrotron DESY
Notkestraße 85, 22607 Hamburg, Germany
jenny.list@desy.de

Published 29 February 2016

The International Linear Collider (ILC) is a proposed e^+e^- collider, focused on precision measurement of the Standard Model and new physics beyond. Polarized beams are a key element of the ILC physics program. The physics studies are accompanied by an extensive R&D program for the creation of the polarized beams and the measurement of their polarization. This contribution will review the advantages of using beam polarization and its technical aspects and realization, such as the creation of polarized beams and the measurement of the polarization.

Keywords: ILC Coll; electron positron: annihilation; polarized beam; accelerator: design.

PACS numbers: 13.88.+e; 29.20.Ej; 29.27.Hj.

1. Introduction

The Standard Model of particle physics is currently regarded the best description of electromagnetic, weak and strong interactions up to the energies investigated during the last decades. Numerous particle physics experiments have confirmed Standard Model predictions, including the discovery of the Higgs boson by the ATLAS and CMS experiments at the Large Hadron Collider (LHC).[1,2] However, there are several unresolved questions hinting that the current formulation of the Standard Model can not be the final fundamental theory. The LHC will continue to search for possible evidence of physics beyond the Standard Model. Since the protons colliding at the LHC are composite particles, the initial state depends on the dynamics of the gluons and quarks in the proton. A lepton machine, where the initial state conditions are very well known, would be an ideal tool to follow up potential discoveries at the LHC with precision measurements and, independent of LHC findings, perform indirect searches for new physics via high precision studies of Standard Model particles and interactions.

The most realistic proposal for a future e^+e^- collider is the International Linear Collider (ILC, see Fig. 1a). The ILC is planned as a linear accelerator with tunable center-of-mass energy of up to about $\sqrt{s} = 500\,GeV$, with the possibility to extend the reach up to $1\,TeV$. Recently, Japan has voiced a strong interest to host the ILC. In June 2013 a Technical Design Report was published.[3–7]

2. The Physics Case for Beam Polarization

The ILC physics program comprises a large spectrum from precision measurements of Standard Model physics and investigations of the Higgs sector to searches for and possible studies of physics beyond the Standard Model. The use of highly polarized beams has many benefits in Standard Model precision tests as well as in the search for new particles and the measurement of their interactions. Longitudinal beam polarization is part of the ILC baseline design, foreseeing a longitudinal polarization of 80% for the electron beam and 30% for the positron beam (with the possibility to upgrade to 60%).

Polarizing both beams leads to four different possible combinations ($e_L^- e_R^+$, $e_R^- e_L^+$, $e_L^- e_L^+$, $e_R^- e_R^+$). In the Standard Model process of e^+e^- annihilation into a vector boson, an electron annihilates a positron of the opposite helicity ($e_L^- e_R^+$ and $e_R^- e_L^+$). The use of polarized beams with opposite signs for electron and positron polarization can therefore be used to enhance the collision cross-section for these processes. In searches for new physics with scalar particles as propagator in the s-channel, same-sign beam polarization can be used to suppress Standard Model background. For example in searches for pair production of dark matter particles in association with a photon, backgrounds can be reduced two orders of magnitude, leading to a significantly enhanced signature.[11]

In addition, asymmetries between the different polarization configurations can be used as an observable to study the properties of the final state particles, such as the chiral structure of their couplings, when they are produced via t/u-channel scattering diagrams. In that case, the helicities of the incoming beams are directly coupled to the helicities of the final particles. Dependent on their couplings to e^+/e^-, specific configurations of beam polarization may be preferred.

This contribution presents a few selected examples for physics studies which benefit from the beam polarization. A more extensive discussion of the physics case for polarized beams at the ILC can be found in Refs. 4, 12.

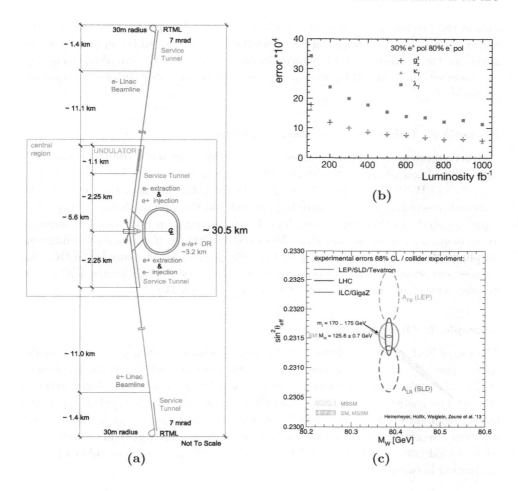

Fig. 1.   **(a)** Schematic layout of the ILC complex for 500 GeV center-of-mass energy, figure courtesy of Ref. 8. **(b)** Absolute uncertainty on triple gauge couplings, figure courtesy of Ref. 9. **(c)** MSSM parameter scan for M_W and $\sin^2 \theta^\ell_{\mathrm{eff}}$. Today's 68% C.L. ellipses from A^b_{FB}(LEP), A^e_{LR}(SLD) and the world average are shown as well as the anticipated LHC and ILC/GigaZ precisions, drawn around today's central value, figure courtesy of Ref. 10.

Example 1: Electroweak couplings of the top quark

Due to its large mass, the top quark is expected to be the Standard Model particle with the strongest coupling to electroweak symmetry breaking mechanisms. Studying the top quark properties can thus be a powerful tool to determine the scale of new physics. At the ILC, the leading order process for top pair production, $e^+e^- \rightarrow t\bar{t}$, gives direct access to the coupling of the top quark to the Z boson and the photon. Unlike the situation at hadron colliders, there is no competing QDC production, so that theoretical uncertainties are greatly reduced. Polarized beams allow to test the chiral structure at the $t\bar{t}Z$ and $t\bar{t}\gamma$ vertices. Using observables such as the production cross-section for left- and right-handed polarized beams, the vector, axial vector and tensorial CP conserving couplings can be separately determined for the Z boson and the photon. In contrast to the LHC, all of the couplings can be resolved at the ILC, with a precision more than one order of magnitude better.[13]

Example 2: Trilinear gauge boson couplings

Another precision test of the electroweak interactions and a possibility to search for signatures of new physics is the study of triple electroweak gauge boson production, e.g. $e^+e^- \to W^+W^-Z$ and $e^+e^- \to ZZZ$. At the ILC, the exact knowledge of the center-of-mass energy of the scattering process and the tunable beam energy allow a precise study of these processes. The beam polarization increases the sensitivity to deviations from the Standard Model and allows to disentangle the different couplings. While these trilinear gauge couplings are also measured at LHC, the foreseen experimental uncertainties are on the 10^{-2} level, while for the ILC a gain in sensitivity of one order or more is expected (see Fig. 1b).[14–16]

Example 3: Model distinction

The use of both polarized electron and positron beams allows a direct probe of the spin of particles produced in resonances. Both a scalar neutrino in the framework of R-parity violating supersymmetry as well as a potential Z' could decay into two muons. The sneutrino as a spin-0 particle couples only to left-handed $e^{\pm}$, so the largest resonance production would occur for the configuration $e_L^- e_L^+$, while in case of a spin-1 resonance such as from a Z' particle, the strongest peak in the resonance curve would come from the $e_L^- e_R^+$ polarization configuration, thus allowing for a distinction between these models.[17]

Example 4: Precision observables

In the past, electroweak precision measurements at the Z resonance were performed at the Large Electron-Positron Collider (LEP) and at the Stanford Linear Collider (SLC). While in general the results are in good agreement, the measurement of the left-right cross-section asymmetry from the SLD experiment at SLC results in slightly different angles for the weak mixing angle from that determined at LEP ($\sin^2\theta_{\text{eff}}(SLC_{A_{\text{LR}}}) = 0.23098 \pm 0.00026$ versus $\sin^2\theta_{\text{eff}}(LEP_{A_{\text{FB}}^b}) = 0.23221 \pm 0.00029$). Whether this difference is a fluctuation or a sign of new physics could be tested by running the ILC at the Z resonance ("GigaZ option"). Via the left-right cross-section asymmetry A_{LR} from all four polarization combinations, defined as

$$A_{LR} = \sqrt{\frac{(\sigma_{RR} + \sigma_{RL} - \sigma_{LR} - \sigma_{LL})(-\sigma_{RR} + \sigma_{RL} - \sigma_{LR} + \sigma_{LL})}{(\sigma_{RR} + \sigma_{RL} + \sigma_{LR} + \sigma_{LL})(-\sigma_{RR} + \sigma_{RL} + \sigma_{LR} - \sigma_{LL})}}, \quad (1)$$

a relative precision of 1.3×10^{-5} can be achieved on $\sin^2 \theta_{\text{eff}}$, which is more than 10 times better than the current precision (see Fig. 1c).[18]

3. Realization at the ILC

The wide range of topics in the ILC physics program which benefit from the use of polarized beams necessitates technologies and strategies for the creation of these polarized beams, their handling as well as the measurement of their polarization. The envisioned concepts for the polarized $e^{\pm}$ sources and the ILC polarimetry concept will be outlined in the following.

3.1. *Polarized sources*

The polarized particle sources at the ILC have to produce bunches with the required beam parameters. In case of the electrons, a high polarization is part of the baseline design for the ILC. The requirements on charge and polarization have already been met in the past at the SLC electron source. The generation of an intense polarized positron beam is more difficult.

3.1.1. *The polarized electron source*

The ILC electron source is located at the beginning of the accelerator complex. It has to produce electron bunches with $>80\%$ longitudinal polarization. The nominal beam parameters require bunch trains of 1312 bunches of 3.0×10^{10} electrons. This beam is produced by a laser-driven photo injector, where circular polarized photons illuminate a photocathode. Stained GaAs/GaAsP superlattice photocathodes yield longitudinal polarized electrons with $\approx 85\%$ polarization and typical quantum efficiency $Q \sim 0.4\%$.

The electron source system comprises two independent laser systems and DC guns. The electrons from either gun are deflected on a beam line, which is equipped with a Mott polarimeter for a measurement of the polarization of the produced electrons, and are pre-accelerated to $5\,GeV$ beam energy before they are injected into the damping rings. The primary challenges for the electron source at the ILC are the development of a laser system for the long bunch trains, and accelerating structures for the normal-conducting part of the pre-accelerator that can handle high RF power. The feasibility of both concepts has been demonstrated in prototypes.[19, 20]

3.1.2. *The polarized positron source*

The positron source uses the high energy electron beam to produce positrons. In the ILC baseline design, the electrons are guided through a $147\,m$ long superconducting helical undulator (with space reserved for upgrades), where they generate multi-MeV circular polarized photons, which in turn produce $e^{\pm}$ pairs with longitudinal polarization upon hitting a thin rotating target made of titanium alloy.

The positrons are captured and also pre-accelerated to $5\,GeV$ before entering the damping ring.

The degree of polarization depends on the undulator parameters and the source design. In the baseline version of the source, the positron bunches are polarized to 30-40%. Collimating the photon beam can increase the average polarization, but reduces the photon yield on the target. With an additional $73.5\,m$ undulator length, a polarization of 60% can be achieved while meeting the requirement of 1.5 positrons per drive-beam electron. The positron yield decreases with lower electron energies. For electron beam energies below $150\,GeV$, the so-called $10\,Hz$ *scheme* has been proposed, alternating a $5\,Hz$ electron beam for the collisions with another $5\,Hz$ electron beam at $150\,GeV$ for the positron source. Recent studies indicate that an electron beam with an energy of $120\,GeV$ could also suffice to achieve an adequate positron yield with a polarization of 30-40%.[21] To measure the polarization at the positron source, a Bhabha polarimeter is proposed.[22] Research and development on some parts of the positron source, such as the design of the target and the photon collimators, is still ongoing.

An electron-driven source is considered as a backup option for the positron beam creation. In this scheme, an electron beam with several GeV is used to generate Bremsstrahlung in a heavy metal target. Like the undulator method, additional development is required, especially to avoid destruction of the target. In case of the electron-driven source, no positron beam polarization would be available.

3.2. *Spin rotators and helicity reversal*

After their creation, the electron and positron beams have emittances that are orders of magnitude too big to reach the desired beam spot sizes in the collisions. To achieve the low emittance required for high luminosities, both beams are temporarily stored in damping rings. To preserve the longitudinal polarization, spin rotators are located before injection into the damping rings to rotate the polarization vector into the vertical beam axis parallel (or anti-parallel) to the magnetic field in the damping rings. After the damping is accomplished, the beams are extracted and

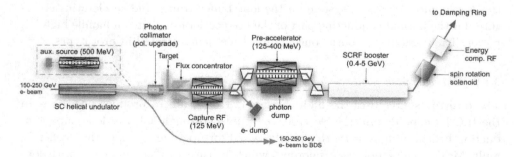

Fig. 2.   Schematic layout of the polarized positron source, figure courtesy of Ref. 23.

transported to the Main Linac systems. During this transport, spin rotators orient the beam polarization to the desired direction.

A fast flipping between the different beam polarization combinations is desirable to reduce the impact of time-dependent variations. The polarization of the electron beam can be flipped easily by reversing the polarity of the laser beam which hits the photocathode at the source. For the positron beam, the direction of the helical undulator winding determines the photon polarization and therefore also the sign of the positron polarization. To switch to the opposite orientation, a dedicated spin flipper is required. One option considered to allow a quick switch between two helicities from train to train at the positron spin rotator upstream of the damping ring is to kick the beam into one of two parallel transport lines with opposite solenoid fields.[24]

3.3. *Polarimetry*

The ILC physics program requires knowledge of the beam polarization with permille-level precision. The overall polarimetry concept at ILC combines the long-term average from e^+e^- collisions with measurements of Compton polarimeters upstream and downstream of the e^+e^- interaction point (IP) (see Fig. 3).[25]

It is planned to instrument both the electron and positron beamline with Compton polarimeters $1.8\,km$ upstream and $160\,m$ downstream of the e^+e^- IP. These Compton polarimeters will be located inside magnetic chicanes, equipped with a laser system that can alternate between left and right circular polarization configurations on a pulse-by-pulse basis. Inside the magnetic chicane, the energy distribution of the Compton-scattered electrons is transformed into a spatial resolution, which is measured using a multi-channel Cherenkov detector. The most precise Compton polarimeter measurement so far, at the Stanford Linear Collider, reached a relative precision of 0.5%.[26] The goal for ILC polarimetry is to improve this at least by a factor of two, i.e. with to systematic uncertainty of 0.25% or better. In or to achieve this, different concepts for Cherenkov detectors and calibration systems are considered.[27-29] The upstream polarimeter, where clean beam conditions and low background rates allow to measure the polarization of each bunch in a bunch train, is capable of an excellent time resolution. At the downstream polarimeter, a

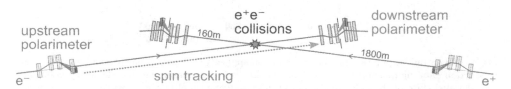

Fig. 3.   Polarimetry concept at the ILC: the luminosity-weighted polarization at the interaction point is determined by extrapolating the Compton polarimeter measurements to the IP. As absolute scale calibration, the long-term average polarization is determined from e^+e^- collision data.

substantially higher laser power is required to overcome the significant background levels present after the collision. While this leads to an operation at a lower sampling rate, it complements the upstream polarimeter by giving access to collision effects.

To relate the measurements at the polarimeter locations to the luminosity-weighted polarization average at the e^+e^- IP, a detailed understanding of both collision effects as well as the evolution of the polarization along the beam delivery system is required. In addition, measurements in the absence of collisions can be used to cross-calibrate the polarimeters if the spin transport between them is well understood. A dedicated software framework for such spin-tracking studies has been developed, showing that it is possible to cross-calibrate the polarimeters to 0.1%, and that both the upstream and the downstream measurements can be individually extrapolated to the e^+e^- IP if the spent beam properties are monitored on a level of 10%.[30]

As absolute scale calibration for the polarization measurement, the long-term average of the polarization can be extracted from e^+e^- collision data by studying any abundant and well-known physics process that is sensitive to the polarization. Several approaches have been studied, e.g. measurements of total cross-sections for various polarization configurations, as well as on single- and double-differential distributions of W pair production.[15,31,32] These methods achieve permille-level precision on the polarization for high integrated luminosities, i.e. on a timescale on the order of years. They rely on an exact helicity reversal. Any changes in the magnitude of the polarization after the helicity reversal have to be corrected based on the polarimeter measurements and their propagation to the IP to reach the precision goal envisaged for polarimetry at the ILC.

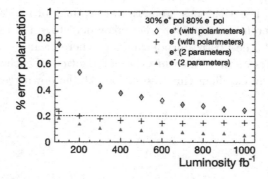

Fig. 4.   Error on the polarization obtained for the application of an angular fit method on W pair production data, figure courtesy of Ref. 33. The idealistic case (assuming that the left-handed and right-handed state of the polarization have the same magnitude) is compared to a more realistic case which takes into account the Compton polarimeter measurements with a precision of 0.25%.

4. Summary

A future lepton collider, such as the proposed International Linear Collider, would be an ideal tool to perform precision measurements of Standard Model physics and the study of phenomena beyond the Standard Model. The publication of the ILC Technical Design report in 2013 as well as the interest of both the Japanese high-energy physics community and the Japanese government to host the ILC in Japan are a positive step towards the realization of the ILC as the next precision collider experiment.

One of the key elements in this endeavor is the use of polarized beams, which offers substantial benefits. The flexible choice of initial state helicities can be used to enhance the effective luminosity, disentangle couplings, suppress backgrounds and, in case signals from physics beyond the Standard Model are found, allow a distinction between new physics models.

To exploit the full potential of the ILC, a high degree of polarization of both beams with the possibility for fast helicity reversal and a precise knowledge of the polarization are required. Designs and techniques for the polarized sources and the polarization measurements exits in a well-advanced state, promising an excellent feasibility to realize the ILC's program for precision measurements.

Acknowledgments

Parts of the presented research has been supported by the BMBF Verbundforschung "Spin Optimierung". Warmest thanks are due to the organizers of the Spin2014 for their successful work ensuring an enjoyable conference experience in Beijing.

References

1. ATLAS Collaboration, *Observation of an Excess of Events in the Search for the Standard Model Higgs boson with the ATLAS detector at the LHC*, Tech. Rep. ATLAS-CONF-2012-093, CERN (Geneva, 2012).
2. CMS Collaboration, *Observation of a new boson with a mass near 125 GeV*, Tech. Rep. CMS-PAS-HIG-12-020, CERN (Geneva, 2012).
3. T. Behnke *et al.*, *arXiv e-prints* (June 2013), arXiv:1306.6327.
4. H. Baer *et al.*, *arXiv e-prints* (June 2013), arXiv:1306.6352.
5. C. Adolphsen *et al.*, *arXiv e-prints* (June 2013), arXiv:1306.6353.
6. C. Adolphsen *et al.*, *arXiv e-prints* (June 2013), arXiv:1306.6328.
7. T. Behnke *et al.*, *arXiv e-prints* (June 2013), arXiv:1306.6329.
8. Figure 2.1 in Ref. 6.
9. Figure 5.22 in Ref. 15.
10. Heinemeyer, Hollik, Weiglein, Zeune *et al.* '13, published as Figure 1.11 in Ref. 18.
11. C. Bartels *et al.*, *Eur.Phys.J.* **C72**, p. 2213 (2012).
12. G. Moortgat-Pick *et al.*, *Physics Reports* **460**, 131 (2008).
13. M. Amjad *et al.*, *ArXiv e-prints* (2013), 1307.8102.
14. M. Schott and J. Zhu, *International Journal of Modern Physics A* **29**, p. 30053 (October 2014).

15. I. Marchesini, Triple gauge couplings and polarization at the ILC and leakage in a highly granular calorimeter, PhD thesis, Universität Hamburg 2011. DESY-THESIS-2011-044.

16. A. Freitas *et al.*, *ArXiv e-prints* (July 2013), 1307.3962.

17. G. Moortgat-Pick, *ArXiv e-prints* (September 2005), hep-ph/0509099.

18. M. Baak *et al.*, *ArXiv e-prints* (2013), 1310.6708.

19. S. Backus and Kapteyn-Murnane Laboratories Inc., *Laser Systems Development for the International Linear Collider (ILC) Photoinjector*, Final Report DE-FG02-06ER84469, US DOE SBIR/STTR Grant (2011).

20. J. Wang *et al.*, Positron Injector Accelerator and RF System for the ILC, in *APAC 2007: Asian Particle Accelerator Conference*, (Indore, India, 2007). SLAC-PUB-12412.

21. A. Ushakov *et al.*, *Simulations of the ILC positron source with 120 GeV electron drive beam*, Tech. Rep. LC-REP-2013-019, Linear Collider Note (2013).

22. G. Alexander *et al.*, *Pramana* **69**, p. 1171 (December 2007).

23. Figure 5.4 in Ref. 6.

24. L. Malysheva *et al.*, *The Spin rotator with a possibility of helicity switching for polarized positron at the ILC*, Tech. Rep. LC-REP-2013-016, Linear Collider Note (2013).

25. B. Aurand *et al.*, *ArXiv e-prints* (August 2008).

26. The ALEPH, DELPHI, L3, OPAL and SLD Collaborations, *Physics Reports* **427**, 257 (2006).

27. C. Bartels *et al.*, *Journal of Instrumentation* **7**, p. 1019 (January 2012).

28. B. Vormwald, From Neutrino Physics to Beam Polarisation - a High Precision Story at the ILC, PhD thesis, Universität Hamburg 2014. DESY-THESIS-2014-006.

29. A. Vauth, A Quartz Cherenkov Detector for Polarimetry at the ILC, PhD thesis, Universität Hamburg 2014. DESY-THESIS-2014-022.

30. M. Beckmann *et al.*, *Journal of Instrumentation* **9**, p. P07003 (2014).

31. K. Mönig, *The Use of positron polarization for precision measurements*, Tech. Rep. LC-PHSM-2000-059, Linear Collider Note (2000).

32. J. List, *PoS* **EPS-HEP2013**, p. 233 (2013).

33. Figure 5.23 in Ref. 15.

Spin Physics (SPIN2014)
International Journal of Modern Physics: Conference Series
Vol. 40 (2016) 1660004 (13 pages)
© The Author(s)
DOI: 10.1142/S2010194516600041

Physics with Polarized Targets in Storage Ring

Toporkov Dmitriy

Budker Institute of Nuclear Physics,
Novosibirsk State University,
Novosibirsk 630090, Russia
D.K.Toporkov@inp.nsk.su

Published 29 February 2016

The method of experiments with internal targets in the storage ring was proposed at the moment of creation of the first storage ring of charged particles. This method has demonstrated dramatic advances during the last 50 years. The most clearly visible progress can be seen in experiments with polarized gas targets. In this report we shall review: (a) the experiments on nuclear physics in the storage rings on the measure of polarization observables in different reactions with electron/positron and proton/deuteron beams, (b) the use of polarized gas targets in the storage ring for polarimetry of the circulated beams. The present status of the activity in this field of experimental physics is given.

Keywords: Storage ring of charged particles; polarized target.

PACS numbers: 13.40.Gp, 29.27.Hj, 29.25.Pj, 29.20.db

1. Introduction

The method of experiments with internal targets in the storage ring of charged particles was proposed at the moment of creation of the first storage rings. This method was first realized in series of experiments which were conducted at the VEPP-2 electron storage ring to study the properties of light nuclei with coincidence detection of the scattered electron and nuclear decay products, including the slow particles in the late 1960s at the Institute of Nuclear Physics, Novosibirsk.[1,2] Soon afterward this method was widely employed in many accelerator laboratories over the world. Almost all advantages of the internal target technique were realized during the succeeding period. The most clearly visible success can be seen in experiments with internal polarized gas targets. The marvelous results have been achieved in the experiments on the measurement of all deuteron electromagnetic form factors, electric form factor of the neutron, on the measurements of the tensor analyzing powers in deuteron photodisintegration. The HERMES experiment at HERA, DESY successfully explored the spin structure of the nucleon. A polarized proton target used as a spin filter at TSR demonstrated a possibility to obtain a polarized circulated proton (in future antiproton) beam. Large number of experiments were performed in ion storage rings with polarized targets and polarized electron-cooled proton or deuteron beams. The goal of these experiments was to measure the analyzing power and spin correlation coefficients in the reactions under the study. A polarized hydrogen jet target was used for the measurement of polarization of circulated protons beam at RHIC. Also a polarized hydrogen jet target having polarized electrons was used at VEPP-3 storage ring to measure the polarization of the circulated electron beam. A list of publications with the use of polarized gas targets in storage rings is so big that it is impossible to mention all of them. I have to omit something, I apologize for that.

2. Experiments with Polarized Targets in the Electron Storage Rings

2.1. *Electromagnetic structure of the deuteron*

The deuteron is the simplest nucleus and the description of its internal structure is a standard test for nuclear theory. In one photon exchange approximation the electromagnetic structure of the deuteron is completely described by the three form factors: the charge monopole, $G_c(Q^2)$, charge quadrupole, $G_Q(Q^2)$, and magnetic dipole, $G_M(Q^2)$, ones, which depend only on the momentum transfer Q squared. The two deuteron structure functions, $A = G_C^2 + 8/9\tau^2 G_Q^2 + 2/3\tau G_M^2$ and $B = 4/3\tau(1 + \tau)G_M^2$, where $\tau = Q^2/4M_d^2$ and M_d is the deuteron mass, can be found from a series of measurements of the differential cross sections $d\sigma/d\Omega$

$$\frac{d\sigma}{d\Omega} = \frac{d\sigma_0}{d\Omega}\{A + B\tan^2(\theta_e/2)\} \tag{1}$$

for elastic unpolarized *ed* scattering under various kinematic conditions and the same Q^2, θ_e — is a scattering angle.[3,4] To separate the charge monopole and quadrupole form factors requires at least one more independent measurement. This additional measurement measurement can be achieved with a tensor-polarized deuterium target.

In such a case the polarization observables (analyzing powers) of the reaction of the elastic scattering are dependent on the form factors of the deuteron and all of them can be extracted separately. The first use of internal tensor polarized deuteron target was realized in 1985 in an electron storage ring VEPP-2.[5] Later similar measurements with internal tensor polarized target in the range of higher momentum transfer were carried out at VEPP-3,[6,7] NIKHEF AmPSM[8,9] and the MIT Bates Linear Accelerator Center.[10] The results on the separated G_c and G_Q are shown in Fig. 1 including the data obtained with fix solid target. The fit of the all data including[7,11] confirms the location of the first node of G_c at $4.19\pm0.05\,\mathrm{fm}^{-1}$, consistent with previous results.

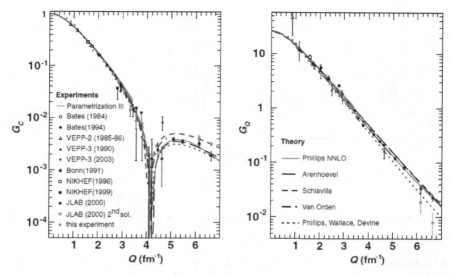

Fig. 1. Results on separated G_C and G_Q compared to various theoretical predictions. From Ref. 10.

2.2. *The Charge form factor of the neutron*

Although the neutron has no net electric charge, it does have a charge distribution. The measurements with thermal neutrons have shown that the neutron has a positive core surrounded by a region of negative charge. The actual distribution is described by the charge form factor G_n^E, which enters the cross section for the elastic electron scattering. Due to the lack of the pure neutron target the information on neutron form factors can be found in experiments with ^{2}H or ^{3}He targets. This is especially true if the targets are polarized. The differential cross section for the $^2\vec{H}(\vec{e}, e'n)p$ reaction with polarized beam and target can be written

$$\frac{d^3\sigma}{d\Omega_e d\Omega_{pq} d\omega} = \sigma_{unp}(1 + \Sigma + P_e\Delta) \tag{2}$$

with $\Sigma = \sqrt{3/2} P_z A_d^V + \sqrt{1/2} P_{zz} A_d^T$, $\Delta = A_e + \sqrt{3/2} P_z A_{ed}^V + \sqrt{1/2} P_{zz} A_{ed}^T$, where σ_{unp} is the unpolarized differential cross section, Pz and Pzz are the vector and tensor polarizations of the deuteron target, P_e is the longitudinal polarization of the electron beam and A_i^j are spin-correlation observables. The beam-target vector polarization observable A_{ed}^V is particularly sensitive to G_n^E in the quasi-elastic neutron-knockout kinematics. For the first time the neutron electric form factor in the reaction $^2\vec{H}(\vec{e}, e'n)p$ were measured at NIKHEF (Amsterdam).[12] Later new measurements of the neutron charge form factor in a wider range of momentum transfer using quasielastic electrodisintegration of the deuteron were performed at the MIT-Bates Linear Accelerator Center.[13] The BLAST detector was used to detect quasielastically scattered electrons in coincidence with recoil neutrons over a range of Q^2 between 0.10 and 0.55 (GeV/c).[2]

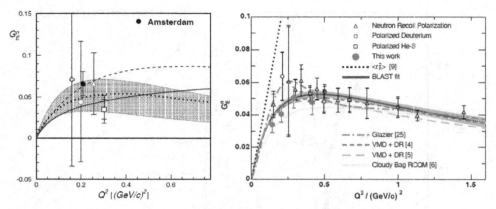

Fig. 2.   World data on G_n^E from double-polarization experiments. Left from Ref. 12, right from Ref. 13.

Also the first precision measurement of the proton electric to magnetic form factor ratio from spin-dependent elastic scattering of longitudinally polarized electrons from a polarized hydrogen internal gas target were performed at MIT-Bates Linear Accelerator Center.[14] The measurement covered the range of four-momentum transfer squared Q^2 from 0.15 to 0.65 (GeV/c).[2]

2.3. *Photodisintegration of polarized deuteron*

One of the most fundamental processes on the deuteron is two-body photodisintegration (PD) $\gamma + d \rightarrow p + n$. It has been a subject of intensive experimental and theoretical research for many years.[15] However several important observables still are measured with insufficient accuracy or not measured at all. The tensor analyzing powers of the reaction accessible through measurement of target asymmetries are among the most poorly known. They are especially interesting because there is a correlation between the degree of tensor polarization and the spatial alignment of the deuteron. Polarization observables are expected to be sensitive to important dynamical details and thus allow in general much more stringent tests of theoretical models. Also the induced recoil proton polarization is consistent with zero above 1 GeV photon energy in disagreement with meson-baryon calculations, see Ref. 15. It's very interesting to investigate if similar effect exists in tensor analyzing powers. A general expression for the cross-section of the two-body PD of the polarized deuteron is written as follows:

$$\frac{d\sigma}{d\Omega} = \frac{d\sigma_0}{d\Omega}\{1 - \sqrt{3/4}P_z \sin\theta_H \sin\varphi_H T_{11} + \sqrt{1/2}P_{zz}\{(3/2\cos^2\theta_H - 1/2)T_{20}$$

$$+ \sqrt{3/8}\sin 2\theta_H \cos\varphi_H T_{21} + \sqrt{3/8}\sin^2\theta_H \cos 2\varphi_H T_{22}\}\} \qquad (3)$$

with σ_0 the unpolarized cross-section, $P_z(P_{zz})$ the degree of vector (tensor) polarization of the target, θ_H the angle between polarization axis and momentum of

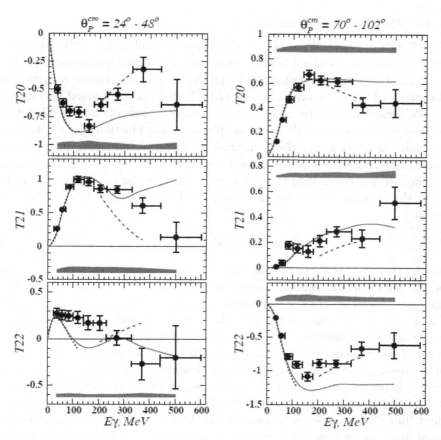

Fig. 3. Tensor analyzing powers vs. photon energy. Vertical bars are statistical uncertainties; horizontal bars indicate the bin size. Shaded bands show systematic uncertainties. Theoretical predictions are from Arenh ovel "N+MEC" (long-dashedline), "N+MEC+IC" (dash-dotted line), and "N+MEC+IC+RC"(solid line) models, from Levchuk (dotted line), and from Schwamb (short-dashed line). From Ref. 19.

γ-quantum, and φ_H the angle between the polarization plane (containing the polarization axis and momentum of the photon) and the reaction plane (containing momenta of the proton and neutron). The tensor analyzing powers T_{2i} are functions of photon energy E_γ and proton emission angle θ_p^{cm}. The measurement of tensor analyzing powers T_{20}, T_{21} and T_{22} in deuteron PD substantially enhanced the quality and span of the existing experimental data. Only few measurements of tensor polarization observables in deuteron PD have been reported up to now.[16−19] The results of these measurements enable an accurate test of available models. Theoretical calculations provide an excellent description of these polarization data below pion production threshold, while above pion production threshold a very good description of T_{20} and T_{22} is demonstrated by a novel approach incorporating a π-MEC retardation mechanism.

2.4. *The HERMES experiment*

The HERMES experiment was located in the East Hall of the HERA facility at DESY. The electron/positron beams automatically became transversely polarized through a Sokolov-Ternov mechanism.[20] The beam spin orientation is rotated into the longitudinal direction just upstream of HERMES, and is rotated back into the transverse direction downstream of the spectrometer. First a polarized ^{3}He internal target was used to measure the neutron spin structure function $g^1(n)$.[21] Later the experiment was investigating the spin structure of the proton and neutron *via* deep-inelastic scattering of polarized positrons/electrons from longitudinally polarized hydrogen and deuterium targets. The nucleon spin can be decomposed conceptually into the angular momentum contributions of its constituents according to the equation

$$\langle s_z^N \rangle = \frac{1}{2} = \frac{1}{2}\Delta\Sigma + Lq + Jg, \tag{4}$$

where the three terms give the contributions to the nucleon spin from the quark spins, the quark orbital angular momentum, and the total angular momentum of the gluons, respectively. In the measurement a value of $\Delta\Sigma = 0.347 \pm 0.024 \pm 0.066$ was obtained[22] in agreement with the result of the Spin Muon Collaboration within their combined uncertainties.[23] Up to today it is still a mystery about what makes the spin of the nucleon. In 1998 the target was converted to one of longitudinally polarized deuterium with nuclear spin one. This allowed not only vector polarization P_z but also tensor polarization P_{zz} of the target to be produced. The latter is related to the structure function b_1 of the deuteron, which HERMES measured for the first time. This pioneer measurement has shown an anomalously large negative value of b_1 in the region $0.2 < x < 0.5$, although the models predict a small or vanishing value of b_1 at moderate x.[24] For the next phase, from 2001 to 2005, a transversely polarized hydrogen target was required to study transversity, the last missing leading-twist structure function of the nucleon.[25] A huge statistical data collected during HERMES experiment is under analysis up to now, see e.g. Ref. 26. A dose of reports on the latest results of the analysis of the data were presented at this Symposium by A. Rostomyan, H. Marukyan and others.

3. Experiments with Polarized Targets in the Ion Storage Rings

3.1. *The FILTEX experiment*

The first use of an internal polarized hydrogen target in the ion ring was performed in 1992 at TSR.[27] Hereafter a polarized proton target used as a spin filter (FILTer EXperiment) at TSR demonstrated a possibility to obtain a polarized circulated proton beam being initially unpolarized.[28] Spin filtering using the spin-dependent part of the nucleon-nucleon interaction is the only experimentally demonstrated viable method to polarize antiprotons. This method is opening a door for new experiments at Facility for Antiproton and Ion Research (FAIR) at GSI, Darmstadt

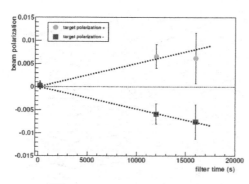 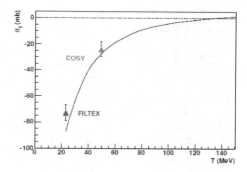

Fig. 4. Left panel: Polarization induced in the beam after filtering for different times and different signs of the target polarization. Right panel: Measured spin-dependent polarizing cross section for the interaction (only statistical errors are shown). From Ref. 29.

aimed to studying hadron structure in the interaction of polarized protons with polarized antiprotons. In 2011 the Polarized Antiproton eXperiments (PAX) Collaboration performed a successful spin-filtering test using protons at Tp = 49.3MeV at the COSY ring, which confirms that spin filtering is a viable method to polarize a stored beam and that the present interpretation of the mechanism in terms of the proton-proton interaction is correct.[29] The results of the experiment are shown in Fig. 4.

3.2. *Nuclear physics experiments*

The very unique possibilities to study nucleons interaction open when both circulated ion beam and target are polarized. An impressive program of such kind of experiments were carried out by the PINTEX group for studying proton-proton and proton-deuteron scattering and reactions between 100 and 500 MeV at the Indiana University Cyclotron Facility (IUCF).[30] An electron-cooled polarized proton or deuteron beams and polarized targets of hydrogen or deuterium were used. The experiments on pp elastic scattering, pion production and to study the three-nucleon force were performed with high accuracy. The quality of the experiment[31] on the elastic pp scattering is illustrated in Fig. 5, right panel.

The other scientific center where similar experiments are in progress right now is a COSY — Cooler Synchrotron of the Forschungszentrum Juelich, which operates with cooled polarized and unpolarized protons and deuteron beams up to momenta of 3.7 GeV/c for the experiments with internal polarized proton and deuteron targets (EDDA, ANKE, WASA, PAX). Some results of EDDA Collaboration[32] on the measurement of spin correlation parameters A_{NN}, A_{SS}, and A_{SL} at 2.1 GeV in pp elastic scattering are shown in Fig. 6. A wide program of scientific research with polarized targets and recent results of the activity at COSY were reported at Symposium by A. Kacharava, G. Ciullo and D. Eversheim.

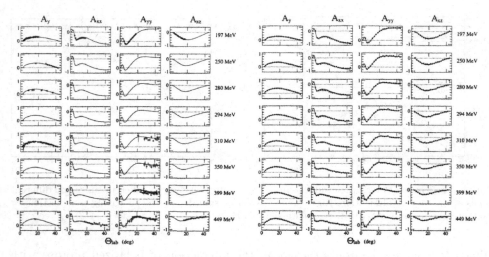

Fig. 5. Analyzing power and spin correlation coefficients as function of energy and angle. The curves are the SM97 phase shift analysis. Left panel — all previously existing data between 175 MeV and 475 MeV from the SAID data base. From Ref. 31.

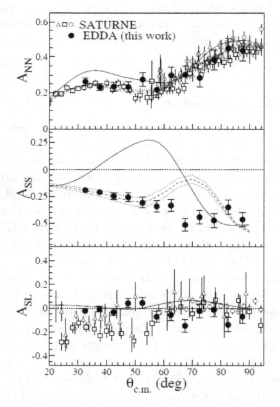

Fig. 6. Results for A_{NN}, A_{SS}, and A_{SL} in comparison to PSA predictions. From Ref. 32.

4. Polarized Target for Polarimetry of the Circulated Beam

4.1. *Polarimetry of the electron beam*

The measurement of the beam polarization in the storage rings and colliders is one of the important tasks in high-energy physics experiments. The method for measurement of an absolute extent of beam polarization in electron-positron storage rings based on MØller scattering on the internal polarized gas target was developed for the first time and successfully applied at VEPP-3 storage ring.[33] A jet of deuterium atoms having atomic electrons totally polarized was used as a polarized electron target. The polarization of the electron beam was determined through the asymmetry measurement

$$A_{NN} = \frac{N_{\uparrow\uparrow} - N_{\uparrow\downarrow}}{N_{\uparrow\uparrow} + N_{\uparrow\downarrow}} = A_g \left| \vec{\xi_B} \right| \left| \vec{\xi_T} \right|. \tag{5}$$

Here $N_{\uparrow\uparrow}$ and $N_{\uparrow\downarrow}$ are the numbers of scattering events registered in two states of the relative polarization orientation provided by switching a sign of target polarization, the factor A_g is less than 1/9 because of a non-zero angular acceptance of the detector. $\vec{\xi_B}$ and $\vec{\xi_T}$ are the polarization vectors of the beam and the target. The measured data serve as a basis for adjusting a scenario of obtaining the polarized beams at VEPP-4M in the energy region near tau-lepton production threshold.

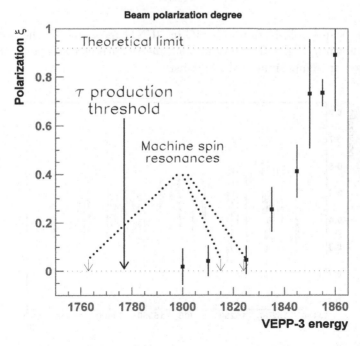

Fig. 7.   Measured electron beam polarization degree vs. VEPP-3 beam energy, from Ref. 33.

4.2. *Polarimetry of the proton beam*

The other application of the polarized atomic hydrogen gas jet is for absolute polarization measurement of the circulated protons beams at RHIC.[34] The measurement is based on the elastic proton-proton scattering in the Coulomb-Nuclear Interference (CNI) region. A_N is a measure of the left–right asymmetry of the cross section in the scattering plane normal to the beam or target polarization. The pp elastic scattering process is 2-body exclusive scattering with identical particles. A_N for the target polarization and the beam polarization should be the same as shown in Equation (6)

$$A_N = -\frac{\varepsilon_{target}}{P_{target}} = \frac{\varepsilon_{beam}}{P_{beam}}, \tag{6}$$

ε_{target} is raw asymmetry for the pp elastic scattering for the transversely polarized proton target and P_{target} is a polarization of the target. The first precise measurement of the analyzing power A_N in pp elastic scattering in the CNI region was performed at RHIC with polarized hydrogen jet the proton polarization of which was well known.[35] The analyzing power A_N in pp elastic scattering in the CNI region was measured to be big enough and weakly depend on the energy for a wide range from 24 GeV to 250 GeV. The beam polarization can be found from the equation

$$P_{beam} = -P_{target}\frac{\varepsilon_{beam}}{\varepsilon_{target}}. \tag{7}$$

The average beams polarization in RHIC in RUN11 measured by the polarized jet target[36] is shown in Fig. 8. The proton polarimetry current status and future plans were reported at Symposium by Y. Makdisi.

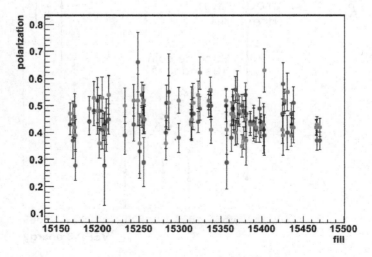

Fig. 8.   H-jet polarization measurements, Run 2011. From Ref. 36.

5. Electromagnetic Form Factors of the Proton

Elastic electron-nucleon scattering is one of the experimental tool for accessing the information on the internal structure of the nucleon. Two structure functions, form factors, $G_E(Q^2)$ and $G_M(Q^2)$ describe the internal structure of the proton. For a long time, the only experimental method to measure form factors was the Rosenbluth method or longitudinal-transverse separation, based on the well-known description of the elastic ep scattering in the one-photon exchange approximation. Another method of direct measuring the ratio G_E/G_M, the so-called polarization transfer method, was proposed in the mid of last century,[37] but could be realized only several decades ago. Surprisingly a clear discrepancy was observed between the results obtained by these two methods for the range of $Q^2 \geq 1\,\mathrm{GeV}^2$, see Fig. 9. It was suggested that the one possible explanation of such a discrepancy is the omission of the two-photon exchange (TPE) contribution in the Rosenbluth formulae. Inclusion of the two-photon exchange correction in the Rosenbluth process may noticeably reconcile the results of two methods.[38] There are three new experiments aimed a precise measurement of the TPE contribution to the elastic ep scattering. Two of them (at VEPP-3 storage ring in Novosibirsk, Russia[39] and the OLYMPUS at DESY, in Hamburg, Germany[40]) used internal hydrogen target and electron/positron beams circulated in the rings. The third experiment has been performed in the Jefferson Lab, USA with liquid hydrogen target and detecting the scattered particles by CLAS detector.[41] The experimentally measured quantity is the ratio $R = \sigma(e^+p)/\sigma(e^-p)$ of the elastic $\sigma(e^+p)$ and $\sigma(e^-p)$ scattering cross sections. The desired hard TPE contribution to the reaction $\delta_{2\gamma}$ was determined from R after taking into account the first order radiative corrections. The results presented as the $R_{2\gamma} = (1 - \delta_{2\gamma})/(1 + \delta_{2\gamma})$ and some theoretical predictions for $R_{2\gamma}$ are shown in Fig. 10. The results obtained show evidence of a significant hard TPE

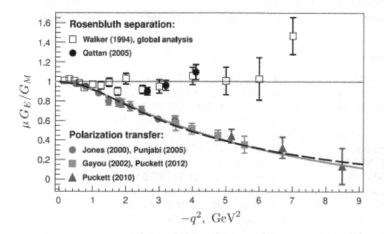

Fig. 9.   The ratio $\mu G_E/G_M$ obtained using different methods.

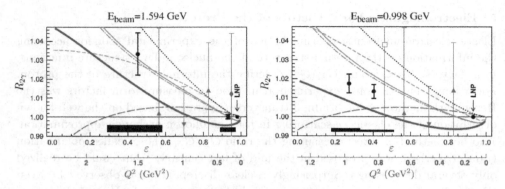

Fig. 10. The data of Novosibirsk experiment (closed circles), earlier measurements and some theoretical predictions (curves) for the ratio $R_{2\gamma}$ as a function of ε or Q^2. From Ref. 39.

effect. They are in moderate agreement with several theoretical predictions of TPE contribution explaining the form factor discrepancy at high Q^2 values.

The situation with the electromagnetic form factors of the proton was discussed at this Symposium in the reports of B. Briscoe, G. Cates and I. Cloet.

6. Conclusion

During the two last decades numerous experiments based on polarized gas targets have been performed in the storage rings of charged particles. Polarized hydrogen, deuterium and ^{3}He gas targets have opened of new possibilities in physical experiments and enable to get data unachievable earlier. A related activity in some laboratories is directed on the preparation of new experiments and creation of more dense polarized targets needed for future experiments. It seems that the ABS technique employing the storage cell for polarized atoms has reached the limit. New ideas being discussed involve the compression of polarized molecules obtaining by nuclear Stern-Gerlach method or made by recombining of polarized atoms.[42,43]

References

1. G. I. Budker *et al.*, *Yadernaya fisika* **6**, 775 (1967).
2. V. F. Dmitriev *et al.*, *Nuclear Physics* **A464**, 237 (1987).
3. R. Gilman and F. Gross, *J. Phys. G* **28**, R37 (2002).
4. M. Garcon and J. W. Van Orden, *Adv. Nucl. Phys.* **26**, 293 (2001).
5. V. F. Dmitriev *et al.*, *Phys. Lett.* **157B**, 143 (1985).
6. R. Gilman *et al.*, *Phys. Rev. Lett.* **65**, 1733 (1990).
7. D. M. Nikolenko *et al.*, *Phys. Rev. Lett.* **90**, 072501 (2003).
8. Ferro-Luzzi *et al.*, *Phys. Rev. Lett.* **77**, 2630 (1996).
9. M. Bouwhuis *et al.*, *Phys. Rev. Lett.* **82**, 3755 (1999).
10. C. Zhang *et al.*, *Phys. Rev. Lett.* **107**, 252501 (2011).
11. D. Abbott *et al.*, *Eur. Phys. J.* **A 7**, 421 (2000).
12. I. Passchier *et al.*, *Phys. Rev. Lett.* **82**, 4988 (1999).
13. E. Geis *et al.*, *Phys. Rev. Lett.* **101**, 042501 (2008).

14. C. B. Crawford *et al.*, *Phys. Rev. Lett.* **98**, 052301 (2007).
15. R. Gilman and F. Gross, *J. Phys. G: Nucl. Part. Phys.* **28**, 37 (2002).
16. M. V. Mostovoy *et al.*, *Phys. Lett.* **B188**, 181 (1987).
17. K. H. Althoff *et al.*, *Z Phys.* **C43**, 375 (1989).
18. S. I. Mishnev *et al.*, *Phys. Lett.* **B302**, 23 (1993).
19. I. A. Rachek *et al.*, *Phys. Rev. Lett.* **98**, 182303 (2007).
20. A. A. Sokolov and I. M. Ternov, *Sov. Phys. Doklady* **8**, 1203 (1964).
21. K. Ackerstaff *et al.*, *Phys. Lett.* **B404**, 383 (1997).
22. A. Airapetian *et al.*, *Phys. Rev.,* **D71**, 012003 (2005).
23. SMC Collaboration, B. Adevaet *et al.*, *Phys. Lett.* **B420**, 180 (1998).
24. A. Airapetian *et al.*, *Phys. Rev. Lett.* **95**, 242001 (2005).
25. A. Airapetian *et al.*, *Phys. Rev. Lett.* **94**, 012002 (2005).
26. http://www-hermes.desy.de/notes/pub/publications.html
27. M. Duren *et al.*, *Nucl. Instrum. Methods* **A 322**, 13 (1992).
28. F. Rathmann *et al.*, *Phys. Rev. Lett.* **71**, 1379 (1993).
29. C. Weidemann for the PAX collaboration. *EPJ Web of Conferences* **66**, 11039 (2014).
30. http://ceem.indiana.edu/documents/Pintex Home Page @ IUCF.pdf
31. B. V. Przewoski *et al.*, *Phys. Rev.* **C 58**, 1897 (1998).
32. F. Bauer *et al.*, *Phys. Rev. Lett.* **90**, 142301 (2003).
33. M. V. Dyug *et al.*, *Nucl. Instrum. Methods* **A 536,** 338 (2005).
34. A. Zelenski *et al.*, *Nucl. Instrum. Method.* **A 536**, 248 (2005).
35. H. Okada *et al.*, *Physics Letters* **B 638**, 450 (2006).
36. A. Zelenski, Polarimetry at RHIC, in *Conf. Proc. PSTP2011*, eds. K. Grigoriev, P. Kravtsov and A. Vasilyev (Russia, St. Petersburg, 2011), p. 39.
37. A. I. Akhiezer and M. P. Rekalo, *Sov. Phys. Dokl.* **13**, 572 (1968).
38. C. E. Carlson and M. Vanderhaeghen, *Annu. Rev. Nucl. Part. Sci.* **57**, 171 (2007).
39. I. A. Rachek *et al.*, arXiv:1411.7372.
40. R. Milner *et al.*, (OLYMPUS Collaboration), *Nucl. Instrum. Methods* **A 741**, 1 (2014).
41. D. Adikaram *et al.*, (CLAS Collaboration), arXiv:1411.6908.
42. D. Toporkov, *Proceedings of Sciences* (PSTP 2013) 064.
43. R. Engels *et al.*, *Rev. Sci. Instrum.* **85**, 103505 (2014).

Spin Physics (SPIN2014)
International Journal of Modern Physics: Conference Series
Vol. 40 (2016) 1660005 (17 pages)
© The Author(s)
DOI: 10.1142/S2010194516600053

World Scientific
www.worldscientific.com

Quark and Glue Components of the Proton Spin from Lattice Calculation

Keh-Fei Liu

On behalf of χQCD Collaboration

Department of Physics and Astronomy, University of Kentucky
Lexington, KY 40506 USA
liu@pa.uky.edu

Published 29 February 2016

The status of lattice calculations of the quark spin, the quark orbital angular momentum, the glue angular momentum and glue spin in the nucleon is summarized. The quark spin calculation is recently carried out from the anomalous Ward identity with chiral fermions and is found to be small mainly due to the large negative anomaly term which is believed to be the source of the 'proton spin crisis'. We also present the first calculation of the glue spin at finite nucleon momenta.

Keywords: Quark spin; glue spin; proton spin.

PACS numbers: 12.38.Gc, 11.40.Ha, 12.38.−t, 14.20.Dh

1. Introduction

Apportioning the spin of the nucleon among its constituents of quarks and glue is one of the most challenging issues in QCD both experimentally and theoretically.

Since the contribution from the quark spin is found to be small ($\sim 25\%$ of the total proton spin) from the global analysis of deep inelastic scattering data,[1] it is expected that the remainder should come from glue spin and the orbital angular momenta of quarks and glue. The quark spin contribution from u, d and s has been studied on the lattice[2,3] since 1995 with quenched approximation or with heavy dynamical fermions.[4] Recently, it has been carried out with light dynamical fermions[5–8] for the strange quark. We will report the calculation of both the connected insertion (CI) and disconnected insertion (DI) contributions to quark spin from u, d, s and c using anomalous Ward identity from the overlap fermion.[11]

As for the quark orbital angular momenta, lattice calculations have been carried out for the connected insertions (CI).[12–19] They are obtained by subtracting the quark spin contributions from those of the quark angular momenta. It has been shown that the contributions from u and d quarks almost cancel each other. Thus for connected insertion, quark orbital angular momenta turn out to be small in the quenched calculation[12, 14] and nearly zero in dynamical fermion calculations.[15–19] On the other hand, gluon helicity distribution $\Delta G(x)/G(x)$ from COMPASS, STAR, HERMES and PHENIX experiments is found to be close to zero.[20–24] A global fit[25] with the inclusion of the polarized deep inelastic scattering (DIS) data from COMPASS[26] and the 2009 data from RHIC,[25] gives a glue contribution $\int_{0.05}^{0.2} \Delta g(x) dx = 0.1 \pm_{0.07}^{0.06}$ to the total proton spin of $1/2\hbar$ with a sizable uncertainty. Most recent analysis[27] of high-statistics 2009 STAR[28] and PHENIX[29] data show an evidence of non-zero glue helicity in the proton. For $Q^2 = 10$ GeV^2, they found the gluon helicity distribution $\Delta g(x, Q^2)$ positive and away from zero in the momentum fraction range $0.05 \leq x \leq 0.2$. However, the result presented in[7] has very large uncertainty in the small x-region. Moreover, it is argued based on analysis of single-spin asymmetry in unpolarized lepton scattering from a transversely polarized nucleon that the glue orbital angular momentum is absent.[30] Given that DIS experiments and quenched lattice calculation thus far reveal that only $\sim 25\%$ of the proton spin comes from the quark spin, lattice calculations of the orbital angular momenta show that the connected insertion (CI) parts have negligible contributions, and gluon helicity from the latest global analysis[27] is $\sim 40\%$ albeit with large error, there are still missing components in the proton spin. In this context, it is dubbed a 'Dark Spin' conundrum.[31, 32]

In this talk, I shall present a complete decomposition of the nucleon spin in terms of the quark spin, the quark orbital angular momentum, and the glue angular momentum in a quenched lattice calculation. I will then summarize the lattice effort in calculating the strange quark spin in dynamical fermions and present a result of the total quark spin from a lattice calculation employing the anomalous Ward identity and, finally, I will show a preliminary first calculation of the glue spin at finite nucleon momenta.

2. Formalism

It is shown by X. Ji[33] that there is a gauge-invariant separation of the proton spin operator into the quark spin, quark orbital angular momentum, and glue angular momentum operators

$$\vec{J}_{\mathrm{QCD}} = \vec{J}_q + \vec{J}_g = \frac{1}{2}\vec{\Sigma}_q + \vec{L}_q + \vec{J}_g, \tag{1}$$

where the quark and glue angular momentum operators are defined from the symmetric energy-momentum tensor

$$J_{q,g}^i = \frac{1}{2}\epsilon^{ijk} \int d^3x \left(T_{q,g}^{0k} x^j - T_{q,g}^{0j} x^k \right), \tag{2}$$

with the explicit expression

$$\vec{J}_q = \frac{1}{2}\vec{\Sigma}_q + \vec{L}_q = \int d^3x \left[\frac{1}{2}\overline{\psi}\,\vec{\gamma}\,\gamma^5\,\psi + \psi^\dagger \left\{ \vec{x} \times (i\vec{D}) \right\} \psi \right], \tag{3}$$

for the quark angular momentum which is the sum of quark spin and orbital angular momentum, and each of which is gauge invariant. The glue angular momentum

$$\vec{J}_g = \int d^3x \left[\vec{x} \times (\vec{E} \times \vec{B}) \right], \tag{4}$$

is also gauge invariant. However, since it is derived from the symmetric energy-momentum tensor in the Belinfante form, it cannot be further divided into the glue spin and orbital angular momentum gauge invariantly.

Since the quark orbital angular momentum and glue angular momentum operators in Eqs. (3) and (4) depends on the radial vector $\vec{r}$, a straight-forward application of the lattice calculation is complicated by the periodic condition of the lattice, and may lead to wrong results.[34] Hence, instead of calculating J_q and J_g directly, we shall calculate them from the energy-momentum form factors in the nucleon.

The Euclidean energy-momentum operators for the quark and glue are

$$T^{(E)}_{\{4i\}q} = (-1)\frac{i}{4}\sum_f \overline{\psi}_f \left[\gamma_4 \overrightarrow{D}_i + \gamma_i \overrightarrow{D}_4 - \gamma_4 \overleftarrow{D}_i - \gamma_i \overleftarrow{D}_4 \right] \psi_f, \tag{5}$$

$$T^{(E)}_{\{4i\}g} = (+i)\left[-\frac{1}{2}\sum_{k=1}^{3} 2\,\mathrm{Tr}^{\mathrm{color}}\left[G_{4k}\,G_{ki} + G_{ik}\,G_{k4} \right] \right]. \tag{6}$$

where we use the Pauli-Sakurai representation for the gamma matrices and the covariant derivative is the point-split lattice operator involving the gauge link U_μ. For the gauge field tensor $G_{\mu\nu}$, we use the overlap fermion Dirac operator. The connection between $G_{\mu\nu}$ and the overlap Dirac operator has been derived[35,36]

$$\mathrm{Tr}_s\left[\sigma_{\mu\nu} D_{\mathrm{ov}}(x,x) \right] = c_T\, a^2\, G_{\mu\nu}(x) + \mathcal{O}(a^3), \tag{7}$$

where Tr_s is the trace over spin. $c_T = 0.11157$ is the proportional constant at the continuum limit for the parameter $\kappa = 0.19$ in the Wilson kernel of the overlap operator which is used in this work. The overlap Dirac operator $D_{\mathrm{ov}}(x,y)$ is exponentially local and the gauge field $G_{\mu\nu}$ as defined in Eq. (7) is chirally smoothed so that it admits good signals for the glue momentum and angular momentum in the lattice calculation.[32]

The form factors for the quark and glue energy-momentum tensor are defined as

$$\langle p', s' | T^{(E)}_{\{4i\}q,g} | p, s \rangle = \left(\frac{1}{2} \right) \bar{u}^{(E)}(p',s') \left[T_1(-q^2)(\gamma_4 \bar{p}_i + \gamma_i \bar{p}_4) \right.$$
$$\left. - \frac{1}{2m} T_2(-q^2)(\bar{p}_4 \sigma_{i\alpha} q_\alpha + \bar{p}_i \sigma_{4\alpha} q_\alpha) - \frac{i}{m} T_3(-q^2) q_4 q_i \right]_{q,g} u^{(E)}(p,s). \tag{8}$$

where the normalization conditions for the nucleon spinors are

$$\bar{u}^{(E)}(p,s)\,u^{(E)}(p,s) \;=\; 1\,, \qquad \sum_s u^{(E)}(p,s)\,\bar{u}^{(E)}(p,s) \;=\; \frac{\not{p}+m}{2m}. \tag{9}$$

2.1. Sum rules and renormalization

The momentum and angular momentum fractions of the quark and glue depend on the renormalization scale and scheme individually, but their sums do not because the total momentum and angular momentum of the nucleon are conserved. We shall use the sum rules as the renormalization conditions on the lattice.

Substituting the energy-momentum tensor matrix elements in Eq. (8) to the matrix elements which define the angular momentum in Eq. (2) and a similar equation for the momentum, it is shown[33] that

$$J_{q,g} = \frac{1}{2} Z^L_{q,g}\, [T_1(0) + T_2(0)]_{q,g}\,, \tag{10}$$

$$\langle x \rangle_{q,g} = Z^L_{q,g}\, T_1(0)_{q,g}, \tag{11}$$

where $Z^L_{q,g}$ is the renormalization constant for the lattice quark/glue operator. $\langle x \rangle_{q,g}$ is the second moment of the unpolarized parton distribution function which is the momentum fraction carried by the quark or glue inside a nucleon. The other form factor, $T_2(0)_{q,g}$, can be interpreted as the anomalous gravitomagnetic moment in analogy to the anomalous magnetic moment, $F_2(0)$.[37, 38]

Since momentum is always conserved and the nucleon has a total spin of $\frac{1}{2}$, we write the momentum and angular momentum sum rules using Eqs. (1), (10) and (11), as

$$\langle x \rangle_q + \langle x \rangle_g \;=\; Z^L_q T_1(0)_q + Z^L_g T_1(0)_g = 1, \tag{12}$$

$$J_q + J_g \;=\; \frac{1}{2}\left\{ Z^L_q\, [T_1(0) + T_2(0)]_q + Z^L_g\, [T_1(0) + T_2(0)]_g \right\} = \frac{1}{2}. \tag{13}$$

It is interesting to note that from Eqs. (12) and (13), one obtains that the sum of the $T_2(0)$'s for the quarks and glue is zero, i.e.

$$Z^L_q T_2(0)_q + Z^L_g T_2(0)_g = 0. \tag{14}$$

We used these sum rules and the raw lattice results to obtain the lattice renormalization constants Z^L_g and Z^L_g and then use perturbation[39] to calculate the quark-glue mixing and renormalization in order to match to the $\overline{MS}$ scheme at 2 GeV which preserves the sum rules.

2.2. Results of a lattice calculation with quenched approximation

Before we present the lattice results, we should point out that the three-point functions for quarks which are needed to extract the form factors in Eq. (8) have two

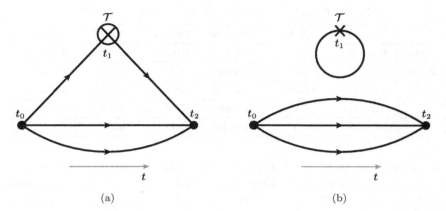

Fig. 1.   Quark line diagrams of the three-point function with current insertion in the Euclidean path integral formalism. (a) Connected insertions (CI), and (b) disconnected insertions (DI).

topologically distinct contributions in the path-integral diagrams: one from connected insertions (CI) and the other from disconnected insertions (DI)[40–43] (See Figs. 1). They arise in different Wick contractions, and it needs to be stressed that they are not Feynman diagrams in perturbation theory. In the case of CI, quark/anti-quark fields from the operator are contracted with the quark/anti-quark fields of the proton interpolating fields. It represent the valence and the higher Fock space contributions from the Z-graphs. In the case of DI, the quark/anti-quark fields from the operator contract themselves to form a current loop, which represents the vacuum polarization of the disconnected sea quarks.

It should be pointed out that, although the quarks lines in the loop and the nucleon propagator appear to be 'disconnected' in Fig 1(b), they are in fact correlated through the gauge background fluctuation. In practice, the uncorrelated part of the loop and the proton propagator is subtracted. The disconnected insertion (DI) refers to the fact that the quark lines are disconnected. For the nucleon, the up and down quarks contribute to both CI and DI, while the strange and charm quarks contribute to the DI only.

A quenched lattice calculation on has been carried out with 3 valence quark masses and extrapolated to the physical pion mass where the numerical details of the calculation are given.[32] We shall present the results in the following table.

For the unrenormalized lattice results, we find that $\left[T_2^u(0) + T_2^d(0)\right]$ (CI) is positive and $T_2^g(0)$ negative, so that the total sum including the small $\left[T_2^u(0) + T_2^d(0) + T_2^s(0)\right]$ (DI) can be naturally constrained to be zero (See Eq. (14)) with the lattice normalization constants $Z_q^L = 1.05$ and $Z_g^L = 1.05$ close to unity. As discussed in Sec. 2.1, the vanishing of the total $T_2(0)$ is the consequence of momentum and angular momentum conservation.

The flavor-singlet g_A^0 which is the quark spin contribution to the nucleon has been calculated before on the same lattice.[2] We can subtract it from the total quark

Table 1.   Renormalized results in $\overline{MS}$ scheme at $\mu = 2$ GeV.

	CI(u)	CI(d)	CI(u+d)	DI(u/d)	DI(s)	Glue
$\langle x \rangle$	0.413(38)	0.150(19)	0.565(43)	0.038(7)	0.024(6)	0.334(55)
$T_2(0)$	0.286(108)	-0.220(77)	0.062(21)	-0.002(2)	-0.001(3)	-0.056(51)
$2J$	0.700(123)	-0.069(79)	0.628(49)	0.036(7)	0.023(7)	0.278(75)
g_A	0.91(11)	-0.30(12)	0.62(9)	-0.12(1)	-0.12(1)	–
$2L$	-0.21(16)	0.23(15)	0.01(10)	0.16(1)	0.14(1)	–

angular momentum fraction $2J$ to obtain the orbital angular momentum fraction $2L$ for the quarks. As we see in Table 1, the orbital angular momentum fractions $2L$ for the u and d quarks in the CI have different signs and they add up to zero, i.e. 0.01(10). This is the same pattern which has been seen with dynamical fermion configurations with light quarks which was pointed out in Sec. 1. The large $2L$ for the u/d and s quarks in the DI is due to the fact that g_A^0 in the DI is large and negative, i.e. $-0.12(1)$ for each of the three flavors. All together, the quark orbital angular momentum constitutes a fraction of 0.47(13) of the nucleon spin. The majority of it comes from the DI. The quark spin fraction of the nucleon spin is 0.25(12) and the glue angular momentum contributes a fraction of 0.28(8). We show the quark spin, the quark orbital angular momentum and the glue angular momentum in the pie chart in Fig. 2. The left panel shows the combination of u and d contributions to the orbital angular momentum from the CI and DI separately while the right panel shows the combined (CI and DI) contributions to the orbital angular momentum from the u and d quarks.

Since this calculation is based on a quenched approximation which is known to contain large uncontrolled systematic errors, it is essential to repeat this calculation with dynamical fermions of light quarks and large physical volume.

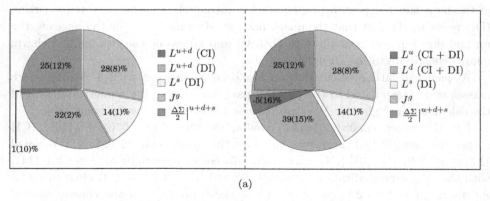

(a)

Fig. 2.   Pie charts for the quark spin, quark orbital angular momentum and gluon angular momentum contributions to the proton spin. The left panel show the quark contributions separately for CI and DI, and the right panel shows the quark contributions for each flavor with CI and DI summed together for u and d quarks.

3. Quark Spin from Anomalous Ward Identity

Attempts have been made to tackle the proton spin decomposition with light dynamical fermions configurations. There have been a number of calculations of the strange quark spin[5–8] which found the strange quark spin Δs to be in the range from -0.02 to -0.03 which is several times smaller than that from a global fit of DIS and semi-inclusive DIS (SIDIS) which gives $\Delta s \approx -0.11$.[1] The large negative contribution from the strange quark is confirmed by a recent analysis[9] of the world data on inclusive deep inelastic scattering data including COMPASS 2010 proton data on the spin asymmetries and the precise JLab CLAS data on the proton and deuteron spin structure functions which gives $\Delta s + \Delta \bar{s} = -0.106 \pm 0.023$.[10]

Such a discrepancy between the global fit of experiments and the lattice calculation of the quark spin from the axial-vector current has raised a concern that the renormalization constant for the flavor-singlet axial-vector current could be substantially different from that of the flavor-octet[44, 45] at the lattice cutoff of ~ 2 GeV. The latter is commonly used for the lattice calculations of the flavor-singlet axial-vector current for the quark spin. To alleviate this concern, we use the anomalous Ward identity (AWI) to calculate the quark spin.[11] The anomalous Ward identity includes a triangle anomaly in the divergence of the flavor-singlet axial-vector current

$$\partial^\mu A^0_\mu = 2 \sum_{f=1}^{N_f} m_f \bar{q}_f i \gamma_5 q_f + i N_f 2q, \tag{15}$$

where q is the local topological charge operator and is equal to $\frac{1}{16\pi^2} G^\alpha_{\mu\nu} \tilde{G}^{\alpha\mu\nu}$ in the continuum. We put this identity between the nucleon states and calculate the matrix element on the right-hand side with a momentum transfer $\vec{q}$ and take the $|\vec{q}| \to 0$ limit

$$\langle p's | A_\mu | ps \rangle s_\mu = \lim_{\vec{q} \to 0} \frac{i|\vec{s}|}{\vec{q} \cdot \vec{s}} \langle p', s | 2 \sum_{f=1}^{N_f} m_f \bar{q}_f i \gamma_5 q_f + 2i N_f q | p, s \rangle. \tag{16}$$

Lattice theory has finally accommodated vector chiral symmetry, the lack of which has hampered the development of chiral fermions on the lattice for many years. It is shown that when the lattice massless Dirac operator satisfies the Gingparg-Wilson relation $\gamma_5 D + D \gamma_5 = a D \gamma_5 D$ with the overlap fermion being an explicit example,[46] the modified chiral transformation leaves the action invariant and gives rise to a chiral Jacobian factor $J = e^{-2i\alpha Tr \gamma_5 (1 - \frac{1}{2} aD)}$ from the fermion determinant.[47] The index theorem[48] shows that this Jacobian factor carries the correct chiral anomaly. It is shown further that the local version of the overlap Dirac operator gives the topological charge density operator in the continuum,[49] i.e.

$$Tr \gamma_5 \left(1 - \frac{1}{2} a D_{ov}(x, x)\right) = \frac{1}{16\pi^2} G^\alpha_{\mu\nu} \tilde{G}^{\alpha\mu\nu}(x) + \mathcal{O}(a) \tag{17}$$

Therefore, Eq. (15) is exact on the lattice for the overlap fermion which gives the correct anomalous Ward identity at the continuum limit. Instead of calculating the matrix element of the axial-vector current derived from the Noether procedure,[48,50] we shall calculate it from the r.h.s. of the AWI in Eq. (15) through the form factors defined in Eq. (16).

In the lattice calculation with the overlap fermion, we note that the renormalization constant of the pseudoscalar density cancels that of the renormalization of the quark mass, i.e. $Z_m Z_P = 1$ for the chiral fermion. Also, the topological charge density, when calculated with the overlap Dirac operator as in the l.h.s of Eq. (17) is renormalized – its integral over the lattice volume is an integer satisfying the Atiya-Singer theorem. Thus, when the matrix elements on the right-hand side of Eq. (16) are calculated with the overlap fermion and its Dirac operator, the flavor-singlet axial-vector current is automatically renormalized on the lattice non-perturbatively *à la* anomalous Ward identity (AWI).

Besides the fact that AWI admits non-perturbative renormalization on the lattice, the pseudoscalar density in DI and the topological density represent the low-frequency and high-frequency parts of the divergence of the axial-vector quark loop respectively. It is learned that on the $24^3 \times 64$ lattice, a mere 20 pairs of the overlap low eigenmodes would saturate more than 90% of the pseudoscalar loop in configurations with zero modes.[51] On the other hand, it is well-known that the contribution to the triangle anomaly comes mainly from the cut-off of the regulator. Therefore, the topological charge density represents the high-frequency contribution of the axial-vector loop, albeit in a local form (the overlap operator is exponentially local). Since the pseudoscalar density is totally dominated by the low modes, we expect that the low-mode averaging (LMA) approach should be adequate for this term. To the extent that the signal for the anomaly term is good, we should be able to calculate the flavor-singlet g_A with the AWI . Both the overlap fermion for the quark loop and the overlap operator for the topological charge density are crucial in this approach.

With the approach described above, we have seen good signals on the $24^3 \times 64$ lattice with the sea quark mass corresponding to a pion mass at 330 MeV.[11] We first show the results for the charm quark which contribute only in the DI. The pseudoscalar density term and the topological charge density term are plotted in Fig. 3(a) as a function of Q^2. We see that the pseudoscalar term is large due to the large charm mass and positive, while the topological charge term is large and negative. When they are added together (black triangles in the figure), it is consistent with zero for the whole range of Q^2. When extrapolated to $Q^2 = 0$, the charm gives zero contribution to the proton spin within error due to the cancellation between the pseudoscalar term and the topological term. It is shown[52] that the leading term in the heavy quark expansion of the quark loop of the pseudoscalar density, i.e. $2mP$ is the topological charge $\frac{2i}{16\pi^2} tr_c G_{\mu\nu} \tilde{G}_{\mu\nu}$, but with a negative sign. Thus, one expects that there is no contribution to the quark spin from heavy quarks to leading order. It appears that the charm quark is heavy enough so that the $\mathcal{O}(1/m^2)$ correction is small. We take this as a cross check of the validity of our

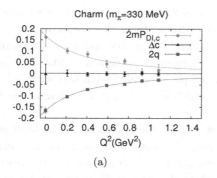
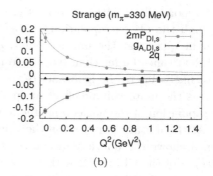

(a)   (b)

Fig. 3. (a) The charm pseudoscalar and topological density contributions to the proton spin as a function of Q^2. (b) The same as in (a) for the strange.

numerical estimate of the DI calculation of the quark loop as well as the anomaly contribution.

The contributions from the strange are also calculated and shown in Fig. 3(b). The $2mP$ contribution is slightly smaller than that of $2q$ and results in a net small negative value for the sum of $2mp$ and $2q$ at finite Q^2. After a dipole fit, we obtain $\Delta s = -0.026(5)$ at $m_\pi = 330$ MeV. Here, Δs denotes the contributions for both s and $\bar{s}$. Δu and Δd are similarly defined in the following.

Since this Δs is quite a bit smaller than the experimental value, we explore the possible finite volume effect and the fact that the induced pseudoscalar form factor $h_A(q^2)$ has been neglected in the Q^2 extrapolation which does not contribute at the $Q^2 = 0$ limit as in Eq. (16), but has a contribution at finite Q^2.[53] We shall check this in the connected insertion (CI) calculation. As can be seen in Fig. 4 for $m_q \sim 500$ MeV, both Δu and Δd in CI calculated from the axial-vector current

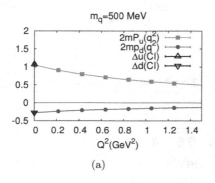
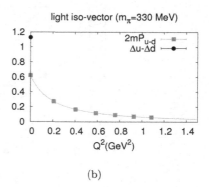

(a)   (b)

Fig. 4. (a) The quark spin of the proton-like baryon with $m_q \sim 500$ MeV from both the axial vector current and the pseudoscalar term through AWI. In this case, the DI contribution of $2mP$ is canceled by the topological charge term. (b) The same as in (a) for light quarks at the unitary point for isovector g_A^3 which involves only CI.

and renormalized with Z_A from the isovector Ward identity are well reproduced through the Q^2 extrapolation of $2mP$ with a dipole form. Whereas, in the case of light quarks at the unitary point, $g_A^3 = 1.13(2)$ from the axial-vector current is 1.8(1) times larger than 0.62(4) from the dipole extrapolation of $2mP$. This is most likely due to the ignorance of the induced pseudoscalar form factor $h_A(q^2)$ as well as the finite volume effect at small Q^2 which is well known to plague the Q^2 extrapolation of the nucleon magnetic form factor.

At the unitary point, when the valence u/d mass matches that of the light sea, Fig. 3 shows the quark spin contribution from the combined pseudoscalar terms $2mP_{u/d}$ of the CI and DI with a dipole extrapolation. Also plotted are the overall quark spin $\Delta u/\Delta d$ by including the topological charge contribution. In this case, we obtain $\Delta u + \Delta d = 0.19(3)$ and $\Delta u - \Delta d = 0.62(4)$ at $Q^2 = 0$ from a dipole extrapolation in Q^2. As we discussed above, the fact that g_A^3 from the axial current is 1.8(1) times larger than that of $\Delta u - \Delta d$ through the Ward identity approach is most likely due to the neglect of the induced pseudoscalar form factor $h_A(q^2)$ and the finite volume effect in the Q^2 extrapolation. We apply this 1.8(1) factor as an estimate to correct the present AWI approach and obtain $\Delta u + \Delta d = 0.35(6)$, $\Delta s = -0.05(1)$. Thus the total estimated spin $\Delta\Sigma = 0.30(6)$ at the unitary point is consistent with the present experimental results which are between 0.2 and 0.3. We expect that, at lighter quark masses, $\Delta\Sigma$ will be smaller.

The above results are from the $24^3 \times 64$ lattice at $m_\pi = 330$ MeV with 200 configurations. The nucleon propagator in the DI has been calculated with the smeared-grid noise source with time dilution which covers all time slices in order to have reasonable statistics for the DI.

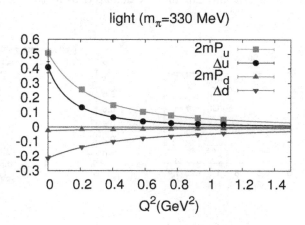

Fig. 5.   The combined pseudoscalar contribution from both the connected insertion (CI) and DI ($2mP_{u/d}$ in the plot), along with the overall quark spin from both pseudoscalar and topological charge ($g_{A,d}$). The plot corresponds to the unitary point with $m_\pi = 330$ MeV.

4. Glue Spin

It has been pointed out that decomposing glue angular momentum into glue spin and orbital angular momentum is only feasible in a specific gauge.[54] Making contact with the parton picture, a spin sum rule involving quark and glue spins and orbital angular momenta is derived in the light-cone gauge (i.e. $A^+ = 0$) with nucleon in the infinite momentum frame.[54] The longitudinal glue spin content is

$$S_G^3 = \langle p, s| \int d^3x Tr(\vec{E} \times \vec{A})^3 |p, s\rangle / \langle p, s|p, s\rangle, \tag{18}$$

where the nucleon state is in the infinite momentum frame and the gauge potential and the gauge field are in the light-cone gauge. Similarly, a gauge-invariant glue helicity distribution is defined with the light-cone correlation function[55]

$$\Delta g(x)S^+ = \frac{i}{2xP^+} \int \frac{d\xi^-}{2\pi} e^{-ixP^+\xi^-} \langle PS|F_a^{+\alpha}(\xi^-)\mathcal{L}^{ab}(\xi^-,0)\tilde{F}_{\alpha,b}^+(0)|PS\rangle, \tag{19}$$

where $\tilde{F}^{\alpha\beta} = (1/2)\,\epsilon^{\alpha\beta\mu\nu}F_{\mu\nu}$ is in the adjoint representation with $\mathcal{A}^+ \equiv T^c A_c^+$, so is the light-cone link $\mathcal{L}(\xi^-,0) = P\exp[-ig\int_0^{\xi^-} \mathcal{A}^+(\eta^-,0_\perp)\,d\eta^-]$.

Since lattice QCD is formulated in Euclidean time, it is not equipped to address the light-cone gauge or the light-cone coordinates and; as such, one is not able to calculate ΔG as defined in Eqs. (18) and (19) on the lattice directly.

On the other hand, a gauge-invariant decomposing of the proton spin has been formulated[56, 57] and examined in various contexts.[58–61] It is based on the canonical energy momentum tensor, instead of that in the symmetric Belinfonte form. The glue spin operator is

$$\vec{S}_g = \vec{E}^a \times \vec{A}_{phys}^a \tag{20}$$

where $A_{\mu\,phys}$ is the physical component of the gauge field A_μ which is decomposed into $A_{\mu\,phys}$ and a pure gauge part as in QED,

$$A_\mu = A_{\mu\,phys} + A_{\mu\,pure}. \tag{21}$$

They transform homogeneously and inhomogeneously with respect to gauge transformation respectively,

$$\begin{aligned} A_{\mu\,phys} &\to A'_{\mu\,phys} = gA_{\mu\,phys}g^{-1} \\ A_{\mu\,pure} &\to A'_{\mu\,pure} = gA_{\mu\,pure}g^{-1} - \frac{i}{g_0}g\partial_\mu g^{-1}, \end{aligned} \tag{22}$$

where g is the gauge transformation matrix and g_0 is the coupling constant. In oder to have a unique solution, conditions are set as follows: the pure gauge part does not give rise to a field tensor by itself and $A_{phys\,\mu}$ satisfies the non-Abelian Coulomb gauge condition

$$F_{\mu\nu\,pure} = \partial_\mu A_{\nu\,pure} - \partial_\nu A_{\mu\,pure} - ig_0[A_{\mu\,pure}, A_{\nu\,pure}] = 0$$

$$D_i A_{i\,phys} = \partial_i A_{i\,phys} - ig_0[A_i, A_{i\,phys}] = 0. \tag{23}$$

This is analogous to the the situation in QED where the photon spin and orbital angular momentum can be defined[62–65] from the canonical energy-momentum tensor

$$\boldsymbol{S}_A = \int \boldsymbol{E}_\perp \times \boldsymbol{A}_\perp \, d^3x, \tag{24}$$

$$\boldsymbol{L}_A = \sum_i \int E_i^\perp \, (\boldsymbol{x} \times \boldsymbol{\nabla}) \, A_i^\perp \, d^3x, \tag{25}$$

where $\perp$ denotes the transverse part. Since they are defined in terms of the transverse parts, they are gauge invariant. However, this gauge invariant definition breaks Lorentz invariance. Nevertheless, it is shown that the 'spin' and 'orbital' angular momentum so defined are conserved for a free field.[63] Furthermore, they are observables and can be measured in experiments through interaction with matter. In 1936, Beth had observed one component of the spin angular momentum of light,[66] by measuring the tongue on a birefringent plate exerted by a circularly polarized light. Also, it is shown[67] that the orbital angular momentum of a paraxial laser beam can be measured. Even though gauge invariance is preserved in this canonical formulation, the spin and orbital AM operators are not boost invariant. Since the experiments are conducted in the lab, the formulation is adequate for this single reference frame.

After integrating the longitudinal momentum x, the light-cone operator for the matrix element has the following expression for the glue helicity[59, 68]

$$H_g = \left[\vec{E}^a(0) \times (\vec{A}^a(0) - \frac{1}{\nabla^+}(\vec{\nabla}A^{+,b})\mathcal{L}^{ba}(\xi^-, 0)) \right]^z. \tag{26}$$

It is recently shown[68] that when boosting the glue spin density operator $\vec{S}_g$ in Eq.(20) to the infinite momentum frame (IMF), the second term in the parentheses on the right side of Eq. (26) is $\vec{A}_{pure}$. Thus H_g is the glue spin density operator $\vec{S}_g$ in the IMF along the direction of the moving frame. In other words, the longitudinal glue spin operator turns into the helicity operator in the IMF.

To carry out a lattice calculation of the matrix element of the glue spin operator, it is realized[69] that $A_{\mu\,phys}$ is related to that fixed in the Coulomb gauge, i.e. $A_{\mu\,phys} = g_c^{-1}A_c g_c$ where A_c is the gauge potential fixed to the Coulomb gauge and g_c is the gauge transformation that fixes the Coulomb gauge. Since $\vec{S}_g$ is traced over color, the spin operator is then

$$\vec{S}_G = \int d^3x \, Tr(g_c \vec{E} g_c^{-1} \times \vec{A}_c) = \int d^3x \, Tr(\vec{E}_c \times \vec{A}_c) \tag{27}$$

where $\vec{E}_c$ is the electric field in the Coulomb gauge. Although it is gauge invariant since both E and A_{phys} transform homogeneouly, it is frame dependent and thus depends on the proton momentum. Its IMF value corresponds to ΔG which is measurable experimentally from high energy proton-proton scattering. The important outcome of the derivation is that glue spin content is amenable to lattice QCD calculation. To the extent that it can be calculated at large enough momentum frame of the proton with enough precision, it can be compared to the experimental glue helicity ΔG.

The first attempt to calculate S_G on the lattice has been carried out on the same set of $2+1$ flavor dynamical domain-wall configurations on the $24^3 \times 64$ lattice with the sea pion mass at 330 MeV.[70] The electric field $\vec{E}$ is constructed from the overlap Dirac operator defined in Eq. (7). The gauge potential $\vec{A}$ is obtained from the unsmeared gauge link. We obtained results for the longitudinal nucleon momenta $p_z = n(2\pi/La)$ with $n = 0, 1, 2$ which correspond to 0, 460 MeV and 920 MeV and for the case of quark masses in the nucleon propagator which correspond to $m_\pi = 380$ MeV and 640 MeV. The unrenormalized results are presented in Fig. 6. We see that the preliminary results in Fig. 6 are quite noisy and, as a result, one cannot discern the p_z behavior. The signal can be improved by smearing the link, but it is a challenge to reach large p_z on the lattice. Since one needs p_z to be less than the cutoff, i.e. $p_z a \ll 1$ to avoid large discretization error, this will require a large lattice size L so that $m_\pi La > 6$ for the nucleon. We note that the question how large a p_z is needed to have the quasi PDFs coincide with the PDFs has been studied in a spectator diquark model.[71] It is found that it is necessary to have p_z as large as 4 GeV for the quasi PDFs to be a good approximation of the PDFs.

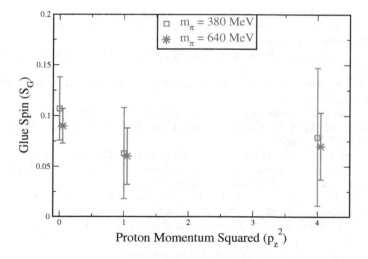

Fig. 6. The results of glue spin S_G in longitudinally polarized proton with longitudinal momenta at 0, 460 MeV and 930 MeV. The quark masses in the nucleon propagator correspond to $m_\pi = 380$ and 640 MeV.

5. Summary

We have reported the current lattice efforts in calculating the quark spin, quark orbital angular momentum, glue angular momentum and glue spin in the nucleon. A complete decomposition of the proton momentum and spin into its quark and glue components is given in a quenched approximation. In this case, the glue angular momentum is not further divided into spin and orbital angular momentum parts. The quark spin calculation is recently carried out from the anomalous Ward identity with chiral fermions and is found to be small mainly due to the large negative anomaly term which is believed to be the culprit of the 'proton spin crisis'. An exploratory lattice calculation of S_G in the non-Abelian Coulomb gauge is carried out[70] which has large errors and the nucleon momentum is limited to ~ 1 GeV. The signal of the glue spin S_G can be improved with smearing, but the major challenge is to have a lattice with fine enough lattice spacing to accommodate large momentum states and show that the infinite momentum extrapolation can be made under control.

Acknowledgments

This work is partially supported by USDOE grant DE-FG05-84ER40154. The author would like to thank X.S. Chen, X. Ji, L. Gamberg, Y. Hatta, E. Leader, C.Lorcé, M. Wakamatsu, and Y. Zhao for helpful and insightful discussions. He also thanks E. Leader and D. Stamenov for providing the strange quark spin contribution from their analysis.

References

1. D. de Florian, R. Sassot, M. Stratmann and W. Vogelsang, Phys. Rev. D **80**, 034030 (2009) [arXiv:0904.3821 [hep-ph]].
2. S. J. Dong, J. -F. Lagae, K. F. Liu, Phys. Rev. Lett. **75**, 2096-2099 (1995), [hep-ph/9502334].
3. M. Fukugita, Y. Kuramashi, M. Okawa and A. Ukawa, Phys. Rev. Lett. **75**, 2092 (1995), [hep-lat/9501010].
4. S. Gusken *et al.* [TXL Collaboration], Phys. Rev. D **59**, 114502 (1999).
5. G. S. Bali *et al.* [QCDSF Collaboration], Phys. Rev. Lett. **108**, 222001 (2012), [arXiv:1112.3354 [hep-lat]].
6. M. Engelhardt, Phys. Rev. D **86**, 114510 (2012) [arXiv:1210.0025 [hep-lat]].
7. A. Abdel-Rehim, C. Alexandrou, M. Constantinou, V. Drach, K. Hadjiyiannakou, K. Jansen, G. Koutsou and A. Vaquero, arXiv:1310.6339 [hep-lat].
8. R. Babich, R. C. Brower, M. A. Clark, G. T. Fleming, J. C. Osborn, C. Rebbi and D. Schaich, Phys. Rev. D **85**, 054510 (2012) [arXiv:1012.0562 [hep-lat]].
9. E. Leader, A. V. Sidorov and D. B. Stamenov, Phys. Rev. D **91**, no. 5, 054017 (2015) [arXiv:1410.1657 [hep-ph]].
10. E. Leader and D. B. Stamenov, private communication.
11. Y. B. Yang, M. Gong, K. F. Liu and M. Sun, PoS LATTICE **2014**, 138 (2014) [arXiv:1504.04052 [hep-ph]].

12. N. Mathur, S. J. Dong, K. F. Liu, L. Mankiewicz, N. C. Mukhopadhyay, Phys. Rev. **D62**, 114504 (2000), [hep-ph/9912289].

13. P. Hagler *et al.* [LHPC and SESAM Collaborations], Phys. Rev. **D68**, 034505 (2003), [hep-lat/0304018].

14. M. Gockeler *et al.* [QCDSF Collaboration], Phys. Rev. Lett. **92**, 042002 (2004) [hep-ph/0304249].

15. D. Brommel *et al.* [QCDSF-UKQCD Collaboration], PoS LATTICE **2007**, 158 (2007), [arXiv:0710.1534 [hep-lat]].

16. J. D. Bratt *et al.* [LHPC Collaboration], Phys. Rev. D **82**, 094502 (2010) [arXiv:1001.3620 [hep-lat]].

17. C. Alexandrou, J. Carbonell, M. Constantinou, P. A. Harraud, P. Guichon, K. Jansen, C. Kallidonis and T. Korzec *et al.*, Phys. Rev. D **83**, 114513 (2011) [arXiv:1104.1600 [hep-lat]].

18. S. N. Syritsyn, J. R. Green, J. W. Negele, A. V. Pochinsky, M. Engelhardt, P. Hagler, B. Musch and W. Schroers, PoS LATTICE **2011** (2011) 178 [arXiv:1111.0718 [hep-lat]].

19. C. Alexandrou, M. Constantinou, S. Dinter, V. Drach, K. Jansen, C. Kallidonis and G. Koutsou, Phys. Rev. D **88**, 014509 (2013) [arXiv:1303.5979 [hep-lat]].

20. C. Adolph *et al.* [COMPASS Collaboration], Phys. Rev. D **87**, 052018 (2013); [arXiv:1211.6849 [hep-ex]]; C. Adolph *et al.* [COMPASS Collaboration], Phys. Lett. B **718**, 922 (2013) [arXiv:1202.4064 [hep-ex]].

21. P. Djawotho [STAR Collaboration], J. Phys. Conf. Ser. **295**, 012061 (2011).

22. A. Airapetian *et al.* [HERMES Collaboration], JHEP **1008**, 130 (2010) [arXiv:1002.3921 [hep-ex]].

23. M. Stolarski [COMPASS Collaboration], Nucl. Phys. Proc. Suppl. **207-208**, 53 (2010).

24. A. Adare *et al.* [PHENIX Collaboration], Phys. Rev. D **79**, 012003 (2009) [arXiv:0810.0701 [hep-ex]].

25. E. C. Aschenauer, A. Bazilevsky, K. Boyle, K. O. Eyser, R. Fatemi, C. Gagliardi, M. Grosse-Perdekamp and J. Lajoie *et al.*, arXiv:1304.0079 [nucl-ex].

26. M. G. Alekseev *et al.* [COMPASS Collaboration], Phys. Lett. B **690**, 466 (2010) [arXiv:1001.4654 [hep-ex]]; Phys. Lett. B **693**, 227 (2010) [arXiv:1007.4061 [hep-ex]].

27. D. de Florian, R. Sassot, M. Stratmann and W. Vogelsang, Phys. Rev. Lett. **113**, no. 1, 012001 (2014) [arXiv:1404.4293 [hep-ph]].

28. L. Adamczyk *et al.* [STAR Collaboration], arXiv:1405.5134 [hep-ex].

29. A. Adare *et al.* [PHENIX Collaboration], Phys. Rev. D **90**, no. 1, 012007 (2014) [arXiv:1402.6296 [hep-ex]].

30. S. J. Brodsky and S. Gardner, Phys. Lett. B **643**, 22 (2006) [hep-ph/0608219].

31. K. F. Liu, M. Deka, T. Doi, Y. B. Yang, B. Chakraborty, Y. Chen, S. J. Dong and T. Draper *et al.*, PoS LATTICE **2011** (2011) 164 [arXiv:1203.6388 [hep-ph]].

32. M. Deka, T. Doi, Y. B. Yang, B. Chakraborty, S. J. Dong, T. Draper, M. Glatzmaier, M. Gong, H.W. Lin, K.F. Liu, D. Mankame, N. Mathur, and T. Streuer, Phys. Rev. D **91**, no. 1, 014505 (2015) [arXiv:1312.4816 [hep-lat]].

33. X. D. Ji, Phys. Rev. Lett. **78**, 610 (1997) [hep-ph/9603249].

34. W. Wilcox, Phys. Rev. D **66**, 017502 (2002) [hep-lat/0204024].

35. K. F. Liu, A. Alexandru and I. Horvath, Phys. Lett. B **659**, 773 (2008) [hep-lat/0703010 [HEP-LAT]].

36. A. Alexandru, I. Horvath and K. F. Liu, Phys. Rev. D **78**, 085002 (2008) [arXiv:0803.2744 [hep-lat]].

37. O. V. Teryaev, hep-ph/9904376.

38. S. J. Brodsky, D. S. Hwang, B. -Q. Ma and I. Schmidt, Nucl. Phys. B **593**, 311 (2001) [hep-th/0003082].
39. M. Glatzmaier and K. F. Liu, arXiv:1403.7211 [hep-lat].
40. K. F. Liu, S. J. Dong, Phys. Rev. Lett. **72**, 1790-1793 (1994), [hep-ph/9306299].
41. K. F. Liu, S. J. Dong, T. Draper, D. Leinweber, J. H. Sloan, W. Wilcox, R. M. Woloshyn, Phys. Rev. **D59**, 112001 (1999), [hep-ph/9806491].
42. K. F. Liu, Phys. Rev. **D62**, 074501 (2000), [hep-ph/9910306].
43. K. -F. Liu, W. -C. Chang, H. -Y. Cheng and J. -C. Peng, Phys. Rev. Lett. **109**, 252002 (2012) [arXiv:1206.4339 [hep-ph]]. %
44. L. H. Karsten and J. Smit, Nucl. Phys. B **183**, 103 (1981).
45. J. F. Lagae and K. F. Liu, Phys. Rev. D **52**, 4042 (1995) [hep-lat/9501007].
46. H. Neuberger, Phys. Lett. B **417**, 141 (1998) [hep-lat/9707022].
47. M. Luscher, Phys. Lett. B **428**, 342 (1998) [hep-lat/9802011].
48. P. Hasenfratz, V. Laliena and F. Niedermayer, Phys. Lett. B **427**, 125 (1998) [hep-lat/9801021].
49. Y. Kikukawa and A. Yamada, Phys. Lett. B **448**, 265 (1999) [hep-lat/9806013]; D. H. Adams, Annals Phys. **296**, 131 (2002) [hep-lat/9812003]; K. Fujikawa, Nucl. Phys. B **546**, 480 (1999) [hep-th/9811235]; H. Suzuki, Prog. Theor. Phys. **102**, 141 (1999) [hep-th/9812019].
50. Y. Kikukawa and A. Yamada, Nucl. Phys. B **547**, 413 (1999) [hep-lat/9808026].
51. M. Gong *et al.* [XQCD Collaboration], Phys. Rev. D **88**, no. 1, 014503 (2013) [arXiv:1304.1194 [hep-ph]].
52. M. Franz, M. V. Polyakov and K. Goeke, Phys. Rev. D **62**, 074024 (2000) [hep-ph/0002240].
53. K. F. Liu, hep-lat/9510046; K. F. Liu, S. J. Dong, T. Draper and W. Wilcox, Phys. Rev. Lett. **74**, 2172 (1995) [hep-lat/9406007].
54. R. L. Jaffe and A. Manohar, Nucl. Phys. B **337**, 509 (1990).
55. A. V. Manohar, Phys. Rev. Lett. **65**, 2511 (1990)
56. X. -S. Chen, X. -F. Lu, W. -M. Sun, F. Wang and T. Goldman, Phys. Rev. Lett. **100**, 232002 (2008) [arXiv:0806.3166 [hep-ph]].
57. X. -S. Chen, W. -M. Sun, X. -F. Lu, F. Wang and T. Goldman, Phys. Rev. Lett. **103**, 062001 (2009) [arXiv:0904.0321 [hep-ph]].
58. M. Wakamatsu, Phys. Rev. D **81**, 114010 (2010) [arXiv:1004.0268 [hep-ph]].
59. Y. Hatta, Phys. Rev. D **84**, 041701 (2011) [arXiv:1101.5989 [hep-ph]].
60. Y. M. Cho, M. -L. Ge and P. Zhang, Mod. Phys. Lett. A **27**, 1230032 (2012) [arXiv:1010.1080 [nucl-th]].
61. E. Leader and C. Lorc, Phys. Rept. **541**, 163 (2014) [arXiv:1309.4235 [hep-ph]].
62. C. Cohen-Tannoudji, J. Dupont-Roc, and G. Grynberg, Photons and Atoms (Wiley, New York 1989).
63. S.J. van Enk and G. Nienhuis, J. Mod. Opt. **41**, 963 (1994); S.J. van Enk and G. Nienhuis, Europhys. Lett. **25**, 497 (1994).
64. K. Y. Bliokh, A. Y. Bekshaev and F. Nori, New J. Phys. **15**, 033026 (2013) [arXiv:1208.4523 [physics.optics]];*ibid* 073022 (2013).
65. K. Y. Bliokh, J. Dressel and F. Nori, New J. Phys. **16**, no. 9, 093037 (2014) [arXiv:1404.5486 [physics.optics]].
66. R.A. Beth, Phys. Rev. **50**, 115 (1936).
67. L. Allen, M.W. Beijersbergen, R.J.C. Spreeuw, and J.P. Woerdman, Phys. Rev. A **45**, 8185 (1992); S.J. van Enk and G. Nienhuis, Opt. Commun. **94**, 147 (1992); M.W. Beijersbergen, L. Allen, , H.E.L.O. van der Veen, and J.P. Woerdman, Opt. Commun. **96**, 123 (1993).

68. X. Ji, J. H. Zhang and Y. Zhao, Phys. Rev. Lett. **111**, 112002 (2013) [arXiv:1304.6708 [hep-ph]].
69. Y.B. Yang and K.F. Liu, under preparation.
70. R. S. Sufian, M. J. Glatzmaier, Y. B. Yang, K. F. Liu and M. Sun, arXiv:1412.7168 [hep-lat].
71. L. Gamberg, Z. B. Kang, I. Vitev and H. Xing, Phys. Lett. B **743**, 112 (2015) [arXiv:1412.3401 [hep-ph]].

Spin Physics (SPIN2014)
International Journal of Modern Physics: Conference Series
Vol. 40 (2016) 1660006 (10 pages)
© The Author(s)
DOI: 10.1142/S2010194516600065

Latest Results from the COMPASS Experiment

M. Stolarski

on behalf of the COMPASS Collaboration

LIP-Lisboa, Avenida Elias Garcia 14
Lisboa, 1000-149, Portugal
mstolars@cern.ch

Published 29 February 2016

In this paper the latest results from the COMPASS experiment are presented. We show results from longitudinally and transversely polarised targets off which high energy muons are scattered. In addition the future plans of COMPASS as well as results of the beam test runs are also presented.

PACS numbers: 13.60.−r, 13.85.−t, 13.88.+e

1. Introduction

COMPASS is a fixed target experiment at CERN SPS accelerator. It has an unique possibility to use both muons and hadron beams of both charges. In this paper a brief summary of the COMPASS spin program is presented. In phase-I of the COMPASS experiment naturally polarized positive muons were impinging on a polarized ^{6}LiD or NH$_3$ target. Results on longitudinal and transverse spin asymmetries from this program are presented in the two next sections. In phase-II, which spin program has just started in the fall of 2014, COMPASS will measure the transverse momentum dependent parton distribution functions in the nucleon as well as its generalized parton distribution functions. Some results from beam tests are presented in section 4. Finally, COMPASS has a rich physics program with hadron beams and unpolarized targets, including: pion polarizability from a Primakoff reaction measurement, exotic particles searches and hadron spectroscopy, to name only a few. These results are not covered here.

2. Longitudinal Physics

In this section recent results on double spin asymmetry A_1^p and the polarized structure function g_1^p are presented. Results of the new COMPASS world g_1 data NLO

QCD fit are also shown. One of the conclusions of this fit is that the gluon polarisation in the nucleon cannot be well constrained even if all the existing inclusive data are used. Therefore, dedicated measurements of the gluon polarization in the nucleon are needed, and such COMPASS results are also presented here.

2.1. *Quark polarization*

From 2002 to 2010 COMPASS used 160 GeV/c muon beam. For the 2011 run the beam energy was increased to 200 GeV/c to access lower values of the Bjorken scaling variable, x, while still in the perturbative region, *i.e.* keeping the negative four momentum transfer Q^2, above 1 (GeV/c)^2. The obtained spin asymmetry A_1^p is presented in the left panel of Fig. 1. Asymmetry is compatible with results from the 2007 run with lower beam energy. Studies of semi-inclusive asymmetries for $h^{\pm}$, and identified $\pi^{\pm}$, $K^{\pm}$ are ongoing.

It is worth adding that COMPASS also studied the low x behaviour of A_1^p for $Q^2 < 1$ (GeV/c)^2. Compared to previous experiments the statistical uncertainties were reduced ten-fold, while a wider domain in x is studied. Even at low values of x asymmetry is found to be positive, around 1%, see the right panel of Fig. 1. The same measurement performed in COMPASS on the deuteron target gave results consistent with zero.[1]

From inclusive asymmetries the g_1^p structure functions were calculated. These new results together with the world $g_1^{p,d,n}$ measurements were used in the new COMPASS NLO fit. The results of fitted singlet Δq_S, Δg and other quark combinations are presented in Fig. 2. Due to the large correlation between Δg and $\Delta \Sigma$ the current world data on inclusive g_1 structure function cannot disentangle between positive and negative gluon polarisation in the nucleon, and related $\Delta \Sigma$ behaviour at low x. Dedicated measurements of $\Delta g/g$ in a semi-inclusive analysis are presented in the next sub-section. The COMPASS g_1^p and g_1^d were also used to test the so called Bjorken Sum Rule. This sum relates the first moment of the non-singlet part of g_1 with a ratio of axial and vector couplings, g_A/g_V. The COMPASS result,

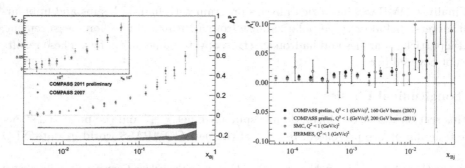

Fig. 1. Spin dependent asymmetry A_1^p, left panel: $Q^2 > 1$ (GeV/c)^2; right panel: $Q^2 < 1$ (GeV/c)^2

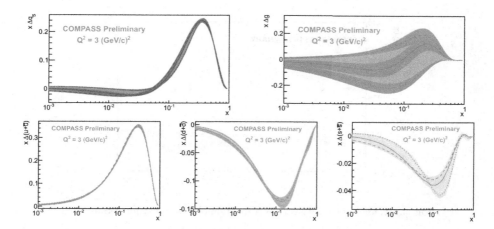

Fig. 2.   Results of COMPASS NLO QCD fit to world g_1 data.

$g_A/g_V = 1.219 \pm 0.052 \pm 0.095$ is in excellent agreement with more precise results from neutron β decay,[2] which gives $g_A/g_V = 1.2701 \pm 0.0025$.

It is worth adding that a low value of the first moment of $\Delta\Sigma$, assuming SU(3) symmetry, leads to large negative polarisation of the strange quarks in the nucleon. The strange polarisation can also be studied in semi-inclusive events with kaon observed in the final state. The strange quark polarization obtained from SIDIS data[3,4] is hardly compatible with the negative expectation from the DIS analysis.

However, the extraction of $\Delta s(x)$ in a SIDIS analysis strongly depends upon the assumed fragmentation functions (FF). In the analyses mentioned before the DSS FF set was used.[5] In case the ratio of fragmentation functions $D_{\bar{s}}^{K^+}/D_u^{K^+}$ is smaller than in DSS, consistent results for inclusive and semi-inclusive analyses. One possibility to study aforementioned FF is to measure kaon multiplicities as a function of x. The sum of charged kaons multiplicities is especially well suited to study $D_{\bar{s}}^{K^+}$. In LO the observed multiplicity is proportional to $\int D_Q^K(z)dz + S/Q \int D_S^K(z)dz$, where $Q = u+\bar{u}+d+\bar{d}$, $S = s+\bar{s}$. Therefore at high x, where S/Q is low, one has an access to $\int D_Q^K(z)dz$ while at lower x, the impact of $S/Q \int D_S^K(z)dz$ should be clearly visible if $D_{\bar{s}}^{K^+}/D_u^{K^+}$ is large. The COMPASS preliminary results are presented in Fig. 3. The expected rise of the multiplicity sum at low x is not seen. COMPASS data seem to prefer lower values of $D_{\bar{s}}^{K^+}/D_u^{K^+}$ than expected from DSS. Finally, COMPASS preliminary results do not agree neither in shape nor in magnitude with HERMES final results.[6] For more details concerning this sub-section refer to two contributions by F. Kunne in these proceedings.

2.2. *Gluon polarisation*

COMPASS has performed several direct gluon extraction measurements.[7–9] Since the leading order photo-absorption process is not sensitive to $\Delta g/g$ one has to study

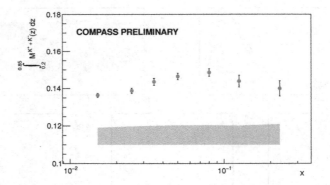

Fig. 3. Kaon multiplicity sum as a function of x, see text for details.

higher order processes like photon-gluon fusion (PGF) to access Δg. For example in LO a clean sample of PGF are events with charmed mesons produced in the final state. The data sample containing high-p_T momentum hadrons is also enriched in PGF events.

Recently, the data used in,[8] were reanalyzed using a new method (all-p_T method). This method allows for the simultaneous extraction of $\Delta g/g$ and the leading process asymmetry A_1 from the same data set. As a result the systematic and statistical errors are reduced comparing to the situation when the values of A_1 asymmetry were taken from as from external source as in.[8]

The preliminary results on $\Delta g/g$ obtained with all-p_T method are presented in the left panel of Fig. 4 in three intervals of gluon momentum fraction x_g. These results are compared with world $\Delta g/g$ obtained in LO analyses. COMPASS data suggest that the $\Delta g/g$ is positive in the measured range of x_g with an average

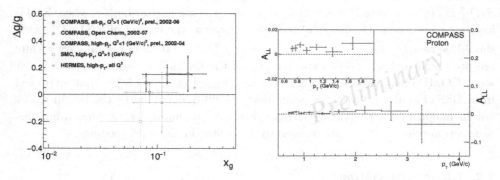

Fig. 4. Left panel: Comparison of the extracted $\Delta g/g$ from the so called all-p_T method with world LO analyses; Right panel: the values of the single inclusive spin dependent asymmetry A_{LL} as a function of hadron p_T.

$\Delta g/g = 0.113 \pm 0.038 \pm 0.035$ at average scale $\mu^2 = 3$ $(\text{GeV}/c)^2$ and average $x_g = 0.10$.

To further constrain $\Delta g/g$ the single inclusive hadron asymmetries were extracted from a low Q^2 data sample, where the hard scale is assured by large transverse momentum of the produced hadron. As an example the extracted asymmetries for the proton target are presented in the right panel of Fig. 4. The obtained results can also be compared with NLO collinear approach calculations.[10] Preliminary COMPASS results are compatible with zero for (not shown here) deuteron target and slightly positive for the proton target. However, in the high p_T region, for proton target theory expects higher asymmetries even in the scenario with maximum positive gluon polarisation. This tension between data and theory is being investigated. For more details concerning COMPASS $\Delta g/g$ measurements see the contribution by K. Kurek in these proceedings.

3. Transverse Physics

Studies of transversely polarised nucleons give access to transverse momentum dependent structure functions which are crucial to understand the 3-D picture of the nucleon.

A known Sivers effect, which is related with parton intrinsic transverse momentum, was recently measured in COMPASS for identified hadrons $\pi^\pm$, $K^{\pm,0}$. In the valence region about 3% effect is observed for π^+ and somewhat larger for K^+. Similarly Collins effect, related to hadron fragmentation p_T, is found to be non-zero for both π^+ and π^-. For the kaon case more data is needed to give a decisive answer. The COMPASS results for both Sivers and Collins effects are presented in left and right panels, respectively, of Fig. 5 as functions of x, p_T and z.[11]

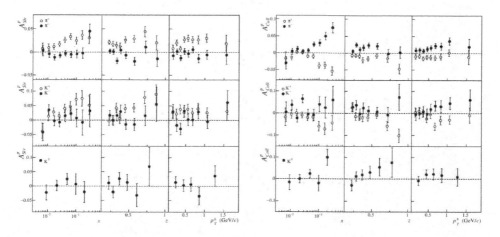

Fig. 5.   Extracted Sivers and Collins asymmetries for identified hadrons, left and right panels respectively.

The Sivers effect measurement were so far mostly related with quarks. In a similar way as in the measurements of $\Delta g/g$, by selecting a sample with high p_T hadrons observed in the final state one can access the gluon Sivers function. Such a measurement was recently performed in COMPASS. The preliminary result based on part of the COMPASS statistics is compatible with zero, $A_{PGF}^{\sin(\phi_{2h}-\phi_S)} = -0.14 \pm 0.15 \pm 0.06$. Results containing all COMPASS statistics are expected soon. For more details see the contribution by A. Szabelski and K. Kurek in these proceedings.

Interesting features are found in studies of two hadron ($2h$) asymmetries, which give access to the interference FF. They were recently measured for identified pairs of $\pi^+\pi^-$, K^+K^-, π^+K^- and $K^+\pi^-$. They are presented in Fig. 6 as a function of x, p_T and invariant mass of the hadron system (M_{inv}). The knowledge of $2h$ asymmetries for proton and deuteron targets allows the extraction of the transversity h_1. The obtained preliminary results are presented in the left panel of Fig. 7 and are compared to a model[12] which uses only single hadron Collins asymmetry. A fair agreement between the extracted transversity from $2h$ analysis and the model is seen. For more details see the contribution by G. Sbrizzai in these proceedings.

Another very interesting study is a comparison of 2h asymmetries with 1h Collins asymmetry, which is presented in the right panel of figure 7. A very striking similarity is observed between the two asymmetries. An important theory achievement is the newly derived cross section formula for the $l^\pm p \to l^\pm + 2h + X$ process, that allows study of asymmetries as a function of the azimuthal angle difference between the two hadrons. Within this formalism one can predict 2h asymmetries from the knowledge of 1h Collins asymmetry. The expected ratio of the two asymmetries

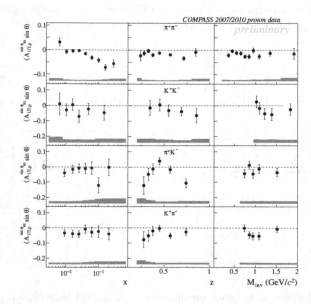

Fig. 6.   Two hadron asymmetry results for pairs of identified hadrons of opposite charge.

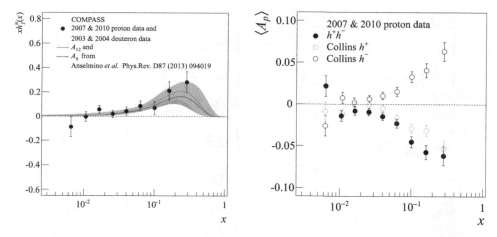

Fig. 7. Left panel: An example of extracted values of transversity $h_1^{u_v}(x)$ and comparison with a model; Right panel: Comparison of $2h$ asymmetries with $1h$ Collins asymmetry.

at high x is about $4/\pi$, in agreement with the experimental measurement. See the contribution by F. Bradamente in these proceedings for more details.

A rapid development of transverse momentum dependent domain asks for multidimensional analyses. These were started in COMPASS and the data are being scrutinized. The example of the bidimensional analysis of the Sivers asymmetry in intervals of Q^2 and x, z, p_T, or the invariant mass of the hadron system, W, is presented in Fig. 8. For more details see the contribution by B. Parsamyan in these proceedings. At the same time COMPASS also analyses hadron multiplicities as functions of x, Q^2, z and most importantly p_T, so that better understanding of intrinsic quark k_T and fragmentation $p_\perp$ can be achieved. For more details see the contribution by N. Makke in these proceedings.

So far mostly Sivers and Collins effects were discussed. However, COMPASS measured also the six other transverse asymmetries which appear in the LO general SIDIS cross-section formula. All but one were found to be consistent with zero. The non zero one $A_{UT}^{\sin \phi_S}$ is related to Sivers $f_{1\perp}$ and transversity h_1.

4. COMPASS Phase-II

In the fall of 2014 COMPASS restarted data taking after two years of shut-down of CERN accelerators due to LHC upgrade. The main goal of the forthcoming 2015 run are studies of transverse momentum dependent structure functions by means of the Drell-Yan process induced by a negative pion beam on a transversely polarised proton target. One of the important expectations is the predicted change of sign of Sivers and Boer-Mulders functions between SIDIS and Drell-Yan processes.

In order to collect a large amount of clean Drell-Yan events, the COMPASS spectrometer had to be modified. A hadron absorber was added directly downstream

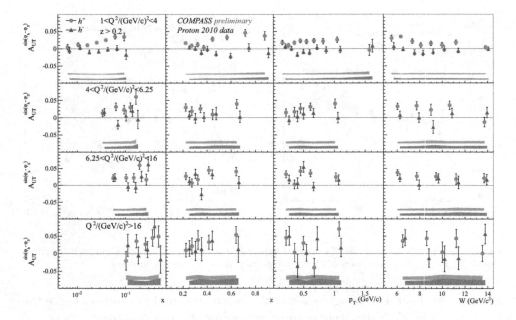

Fig. 8. Measured Sivers effect in bidimensional grid of Q^2 and x, z, p_T, or W.

the target. Although, in this way the multiple-scattering increase one has verified that the mass resolution of the dimuon spectrum is not too much degraded. Moreover, to study polarized effects COMPASS simultaneously collects data with target cells polarised in opposite directions. Therefore it is also mandatory that the vertex resolution along the beam line direction is good enough to separate the two cells.

The results of a beam test performed in 2009 are presented in Fig. 9, being satisfactory in what concerns these two problems. compared to a prototype absorber used in 2009 test, The final one is expected to improve both the mass resolution and the vertex resolution. The vertex resolution will be further reduced by installing a dedicated vertex detector in the middle of the hadron absorber.

COMPASS aims to collect totally 180000 events of Drell-Yan in the high dimuon mass region 4-9 GeV/c^2. The expected statistical errors for various asymmetries are presented in left panel of Fig. 10. An example of the theoretical prediction[13] is shown in the right panel of Fig. 10. If the asymmetries are as large as predicted by some theory groups, COMPASS should observe rather significant non zero results. For more details see the contribution by B. Parsamyan in this proceedings. Details about possible unpolarized Drell-Yan studies using data to be collected and (additional) Al and W targets are presented in paper by W.-C. Chang in these proceedings.

Other COMPASS plans include Generalized Parton Distribution (GPD) measurements, especially in the golden channel of Deeply Virtual Compton Scattering (DVCS). The GPD give also information about the 3-D structure of the nucleon,

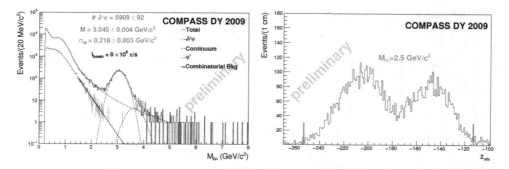

Fig. 9. Left panel: Mass resolution of J/Ψ obtained from a test with hadron absorber, Right panel: separation of interaction vertex between two target cells.

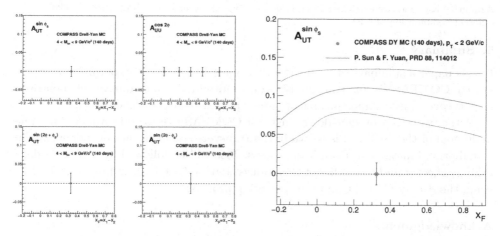

Fig. 10. Left panel: expected statistical errors in each of the relevant azimuthal modulations; Right panel: comparison of the expected precision for the Sivers effect with a model prediction.

with one of the dimension being the impact parameter space. Some of the integrals involving GPD are connected with angular momentum of quarks in the nucleon.

A pilot run data were taken in 2012. The obtained results, seen in Fig. 11, show that indeed an excess of events attributed to DVCS above the Bethe-Heitler (BH) background is observed in high x region, as expected. For more details see the contribution by A. Ferrero in these proceedings.

The generalized parton distributions can be also obtained from studies of hard exclusive vector meson production. As usually various modulations of the cross section are sensitive to different combinations of GPD. Such studies were performed in COMPASS, so far for the ρ meson (studies for ω meson are being finalized). The extracted asymmetries are generally consistent with zero, except for some deviation observed for $A_{UT}^{\sin\phi_s}$. The asymmetries are presented in the right panel of Fig. 11. For more details see the contribution from J. ter Wolbeek in these proceedings.

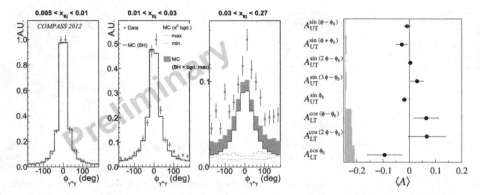

Fig. 11. Left panel: Extracted spectra for events with hard exclusive photon in the final state in three intervals of x. Clear indication of DVCS events is visible in the highest x region. Right Panel: Azimuthal asymmetries extracted from hard exclusive ρ production, each of which is related to a different combination of GPD.

5. Summary

In this paper a summary of recent results from the COMPASS experiment was presented. COMPASS is finalizing analyses on longitudinally and transversely polarised target; several new important results are expected to be published soon.

After two years of shut-down, last fall COMPASS restarted data taking. The main goal of the 2015 run is to study the transverse momentum dependent parton distribution functions in Drell-Yan process. This will be followed by a 2016-17 data taking concentrated on studies of Generalized Parton Distribution functions, mainly using the deeply virtual Compton scattering process.

Acknowledgments

This research was supported by the Portuguese Fundação para a Ciência e a Tecnolagia, grant SFRH/BPD/64853/2009.

References

1. COMPASS Collab. (V.Yu. Alexakhin *et al.*), *Phys. Lett. B* **647**, 330 (2007).
2. Particle Data Group (J. Beringer *et al.*), *Phys. Rev. D* **86**, 010001 (2012).
3. HERMES Collab. (A. Airapetian *et al.*), *Phys Lett. B* **666**, 446 (2008).
4. COMPASS Collab. (M. Alekseev *et al.*), *Phys. Lett. B* **680**, 217 (2009).
5. D. de Florian, R. Sassot and M. Stratmann, *Phys. Rev. D* **75**, 114010 (2007).
6. HERMES Collab. (A. Airapetian *et al.*), *Phys. Rev. D* **87**, 074029, (2013).
7. COMPASS Collab. (E. S. Ageev *et al.*), *Phys. Lett. B* **633**, 25 (2006).
8. COMPASS Collab. (C. Adolph *et al.*), *Phys. Lett. B* **718**, 922, (2013).
9. COMPASS Collab. (C. Adolph *et al.*), *Phys. Rev. D* **87**, 052018, (2013).
10. B. Jaeger, M. Stratmann, and W. Vogelsang, *Eur. Phys. J. C* **44**, 533 (2005).
11. COMPASS Collab. (C. Adolph *et al.*), *hep-ex/1408.4405 subm. to Phys. Lett. B*
12. M. Anselmino *et al.*, *Phys. Rev. D* **87**, 094019 (2013).
13. P. Sun and F. Yuan, *Phys. Rev. D* **88**, 114012 (2013).

Spin Physics (SPIN2014)
International Journal of Modern Physics: Conference Series
Vol. 40 (2016) 1660007 (1 page)
© The Author(s)
DOI: 10.1142/S2010194516600077

Summary Report for PSTP2013

M. Poelker

Center for Injectors and Sources
Thomas Jefferson National Accelerator Facility
12000 Jefferson Ave., Newport News, VA 23606, USA
poelker@jlab.org

D. G. Crabb

Department of Physics, University of Virginia
382 McCormick Rd.
Charlottesville, VA 22903, USA
dgc3q@cms.mail.virginia.edu

Published 29 February 2016

The 15th International Workshop on Polarized Source Targets and Polarimeters was held at the University of Virginia, Charlottesville during September 9–13, 2013. It was sponsored jointly by the University of Virginia, Jefferson Laboratory and the International Spin Physics Committee. A summary of the workshop was presented.

Keywords: Polarized source; polarized targets; polarimeters; PSTP.

The PSTP2013 Workshop at the University of Virginia, Charlottesville continued a long tradition of bringing together experts in the technical fields necessary to carry out particle physics experiments involving spin at accelerators around the world. Eighty one scientists were registered for the Workshop but unfortunately four from Russia and two from China could not attend because of long waiting periods to obtain travel visas. Similar to past workshops, the topics covered were: Polarized Solid and Gas Targets, Polarized Ion and electron (positron) Sources, Polarized Electron and Ion Beam Polarimetry, Applications of Spin, and New Initiatives. PSTP2013 proceedings were published in *Proceedings of Science*, an open-access online journal organized by SISSA, the International School for Advanced Studies based in Trieste (http://pos.sissa.it/cgi-bin/reader/conf.cgi?confid=182). Paper copies are available on request.

Spin Physics (SPIN2014)
International Journal of Modern Physics: Conference Series
Vol. 40 (2016) 1660008 (11 pages)
© The Author(s)
DOI: 10.1142/S2010194516600089

Three Dimensional Imaging of the Nucleon — TMD (Theory and Phenomenology)

Zuo-Tang Liang

*School of Physics and Key Laboratory of Particle Physics & Particle Irradiation (MOE),
Shandong University, Jinan, Shandong 250100, China*

Published 29 February 2016

This is intend to provide an overview of the theory and phenomenology parts of the TMD (Transverse Momentum Dependent parton distribution and fragmentation functions) studies. By comparing with the theoretical framework that we have for the inclusive deep inelastic lepton-nucleon scattering and the one-dimensional imaging of the nucleon, I try to outline what we need to do in order to construct a comprehensive theoretical framework for semi-inclusive reactions and the three dimensional imaging of the nucleon. After that, I try to give an overview of what we have already achieved and make an outlook for the future.

Keywords: TMD parton distribution; twist; collinear expansion; semi-inclusive.

PACS numbers: 12.38.−t, 12.38.Bx, 12.39.St, 13.60.-r, 13.66.Bc, 13.87.Fh, 13.88.+e, 13.40.−f, 13.85.Ni

1. Introduction

With the deep going of the study of the nucleon structure, three dimensional imaging has become the frontier and a hot topic in recent years. It is commonly recognized that the three dimensional imaging contains much more abundant physics on the nucleon structure and the properties of QCD. The study was initially triggered by the experimental finding of striking single-spin asymmetries (SSA) in inclusive hadron production in hadron-hadron collisions. Gradually it grows into a field aiming at a comprehensive picture of nucleon structure including spin and transverse momentum dependences.

The one dimensional imaging of the nucleon is provided by the parton distribution functions such as the number densities, $q(x)$, the helicity distributions, $\Delta q(x)$, and the transversities, $\delta q(x)$, for quarks of different flavors. In the three dimensional case, i.e. where the transverse momentum is also considered, not only the direct extensions of them to include transverse momenta are involved, but also many other correlation functions such as the Sivers function, the Boer-Mulders function, the pretzelocity etc exist. Moreover, higher twist effects become also important and need to be considered consistently.

The study on the three dimensional imaging of the nucleon is in a rapid developing phase and it is not so easy to present a comprehensive overview of all different aspects of studies. Here, I choose to do the job in the following way: First, I will try to make a brief review of what we did in one dimensional case with inclusive DIS. In this way, I hope that I can show you the main line of what we need to do in three dimensional case. Then I will try to summarize what we have already achieved along this line and what we need to do next. For the sake of space, I will mainly concentrate on the discussions but keep as less equations as possible. An extended version is prepared and will be published in a special issue of Frontier of Physics.

2. Inclusive DIS & the One Dimensional Imagining of the Nucleon

Our studies on the structure of a fast moving nucleon started with the inclusive deep inelastic scattering (DIS) e.g. $e^- + N \to e^- + X$. We recall that, under one photon exchange, the differential cross section is given by the Lorentz contraction of the known leptonic tensor and the hadronic tensor $W_{\mu\nu}(q, p, S)$ (where p and S are the 4-momentum and polarization vector of the nucleon). Information of the structure of the nucleon is contained in the hadronic tensor $W_{\mu\nu}(q, p, S)$.

The theoretical framework for inclusive DIS has been constructed in the following steps. First, we studied the kinematics and obtained the general form of the hadronic tensor by applying the basic constraints from the general symmetry requirements such as Lorentz covariance, gauge invariance, parity conservation and Hermiticity. We found out that the hadronic tensor is determined by four independent structure functions F_1, F_2, g_1 and g_2, where the first two describe the unpolarized case and the latter two are needed for polarized cases.

Our knowledge of one dimensional imaging of the nucleon starts with the "intuitive parton model" that is very nicely formulated e.g. in [1]. Here, it was argued that, in a fast moving frame, because of time dilation, quantum fluctuation such as vacuum polarizations, can exist quite long. In the infinite momentum frame, such fluctuations exist for ever. In this case, a fast moving nucleon can be viewed as a beam of free "partons". The probability of the scattering of an electron with the nucleon is equal to that of the scattering with a parton convoluted with the number density of the parton in the nucleon, i.e.,

$$|\mathcal{M}(eN \to eX)|^2 = \sum_q \int dx f_q(x) |\hat{\mathcal{M}}(eq \to eq)|^2, \tag{1}$$

where $f_q(x)$ is the number density of q in the nucleon. In this way, we obtained the famous results, $F_2(x, Q^2) = 2xF_1(x, Q^2) = \sum_q e_q^2 x f_q(x)$, $g_1(x, Q^2) = \sum_q e_q^2 x \Delta f_q(x)$

and so on. Here, I would like to point out that, with this intuitive parton model, we are doing nothing else but the impulse approximation that we often used in describing a collision process where we do the following approximations,

- during the interaction of the electron with the parton, interactions between the partons are neglected;
- the scatterings of the electron with different partons are added incoherently;
- the electron interacts only with one single parton.

Although the intuitive model is elegant and practical, we are not satisfied since it is not easy to control the accuracy. The proper quantum field theoretical (QFT) formulation is given by starting with Feynman digram Fig. 1(a) where we obtain,

$$W_{\mu\nu}^{(0)}(q, p, S) = \frac{1}{2\pi} \int \frac{d^4 k}{(2\pi)^4} \text{Tr}[\hat{H}_{\mu\nu}^{(0)}(k, q)\hat{\phi}^{(0)}(k, p, S)], \qquad (2)$$

where k is the 4-momentum of the parton, $\hat{H}_{\mu\nu}^{(0)}(q, k) = \gamma_\mu(\slashed{k} + \slashed{q})\gamma_\nu(2\pi)\delta_+((k-q)^2)$ is a calculable hard part and the matrix element

$$\hat{\phi}^{(0)}(k, p, S) = \int d^4 z e^{ikz} \langle p, S|\bar{\psi}(0)\psi(z)|p, S\rangle, \qquad (3)$$

is known as the quark-quark correlator describing the structure of the nucleon. By taking the collinear approximation, i.e. taking $k \approx xp$, and neglecting the power suppressed contributions, we obtain exactly the same result as that obtained using Eq. (1) based on the intuitive parton model. [Here, we use light-cone coordinate $k^\mu = (k^+, k^-, \vec{k}_\perp)$ and take $\bar{n} = (1, 0, \vec{0}_\perp)$, $n = (0, 1, \vec{0}_\perp)$, $n_\perp = (0, 0, \vec{n}_\perp)$. Also, we choose the coordinate system such that $p = p^+\bar{n}$.] At the same time, we obtain also a QFT operator expression of $f_q(x)$,

$$f_q(x) = \int \frac{dz^-}{2\pi} e^{ixp^+ z^-} \langle p|\bar{\psi}(0)\frac{\gamma^+}{2}\psi(z)|p\rangle, \qquad (4)$$

which is indeed the number density of parton in the nucleon. However, from this expression, we see immediately that it is not (local) gauge invariant! To get the gauge

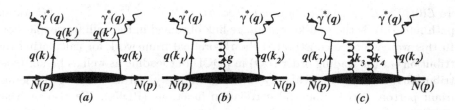

Fig. 1. Examples of the Feynman diagram series with multiple gluon scattering considered for $\gamma^* + N \to q + X$ with (a) $j = 0$, (b) $j = 1$ and (c) $j = 2$ gluons exchanged.

invariant formulation, we need to take into account the multiple gluon scattering given by the diagram series shown in Fig.1(a-c). The contribution from each diagram is expressed as a trace of a hard part and a matrix element. E.g., for Fig. 1(b),

$$W_{\mu\nu}^{(1,L)}(q,p,S) = \frac{1}{2\pi} \int \frac{d^4k_1}{(2\pi)^4} \frac{d^4k_2}{(2\pi)^4} \text{Tr}[\hat{H}_{\mu\nu}^{(1,L)}(k_1,k_2,q)\hat{\phi}_\rho^{(1)}(k_1,k_2,p,S)], \quad (5)$$

$$\hat{\phi}_\rho^{(1)}(k_1,k_2,p,S) = \int d^4z d^4y e^{ik_1 z + (k_2-k_1)y} \langle p,S|\bar{\psi}(0)A_\rho(y)\psi(z)|p,S\rangle, \quad (6)$$

where L in the superscript denotes left cut. The matrix element is now a quark-j-gluon(s)-quark correlator, where j is the number of gluons. We also see that none of such quark-j-gluon(s)-quark correlators is gauge invariant.

To get the gauge invariant form, we need to apply the collinear expansion as proposed in Refs. [2-4], which is carried out in the following four steps.

(1) Make Taylor expansions of all hard parts at $k_i = x_i p$, e.g.,

$$\hat{H}_{\mu\nu}^{(0)}(k,q) = \hat{H}_{\mu\nu}^{(0)}(x) + \frac{\partial\hat{H}_{\mu\nu}^{(0)}(x)}{\partial k_\rho}\omega_\rho^{\rho'}k_{\rho'} + \cdots, \quad (7)$$

where $\omega_\rho^{\rho'}$ is a projection operator defined by $\omega_\rho^{\rho'} \equiv g_\rho^{\rho'} - \bar{n}_\rho n^{\rho'}$.

(2) Decompose the gluon field into $A_\rho(y) = A^+(y)\bar{n}_\rho + \omega_\rho^{\rho'}A_{\rho'}(y)$.

(3) Apply the Ward identities such as,

$$\frac{\partial\hat{H}_{\mu\nu}^{(0)}(x)}{\partial k_\rho} = -\hat{H}_{\mu\nu}^{(1)\rho}(x,x), \qquad p_\rho\hat{H}_{\mu\nu}^{(1,L)\rho}(x_1,x_2) = \frac{H_{\mu\nu}^{(0)}(x_1)}{x_2 - x_1 - i\epsilon}. \quad (8)$$

(4) Add all terms with the same hard part together and we obtain $W_{\mu\nu}(q,p,S) = \sum_{j,c} \tilde{W}_{\mu\nu}^{(j,c)}(q,p,S)$ ($j = 0,1,2,\cdots$ and c is the different cut) where e.g. for $j = 0$,

$$\tilde{W}_{\mu\nu}^{(0)}(q,p,S) = \frac{1}{2\pi} \int \frac{d^4k}{(2\pi)^4} \text{Tr}[\hat{H}_{\mu\nu}^{(0)}(x)\ \hat{\Phi}^{(0)}(k,p,S)], \quad (9)$$

$$\hat{\Phi}^{(0)}(k,p,S) = \int d^4y e^{iky} \langle p,S|\bar{\psi}(0)\mathcal{L}(0;y)\psi(y)|p,S\rangle, \quad (10)$$

where $\mathcal{L}(0;y) = \mathcal{L}^\dagger(\infty;0)\mathcal{L}(\infty;y)$, and $\mathcal{L}(\infty;y) = Pe^{-ig\int_{y^-}^\infty d\xi^- A^+(\xi^-,\vec{y}_\perp)}$ (P stands for path integral), is the well-known gauge link obtained in the collinear expansion.

In this way, we have constructed the theoretical framework for calculating the contributions at the leading order (LO) in pQCD but leading as well as higher twist contributions in a systematical way. The results are given in terms of the gauge invariant parton distribution and correlation functions (PDFs). We also see that the PDFs involved here are all scale independent. This is because we have till now considered only the LO pQCD contributions, i.e. the tree diagrams.

Because the hard parts in $\tilde{W}_{\mu\nu}^{(j)}$'s such as that given by Eq. (9) are only functions of the longitudinal component x but independent of other components of the parton momentum k, we can simplify them. E.g., for $j = 0$, it reduces to,

$$\tilde{W}_{\mu\nu}^{(0)}(q,p,S) = \frac{1}{2\pi} \int p^+ dx \text{Tr}\left[\hat{H}_{\mu\nu}^{(0)}(x) \ \hat{\Phi}^{(0)}(x,p,S)\right], \tag{11}$$

$$\hat{\Phi}^{(0)}(x,p,S) = \int \frac{dy^-}{2\pi} e^{ixp^+y^-} \langle p,S|\bar{\psi}(0)\mathcal{L}(0;y^-)\psi(y^-)|p,S\rangle. \tag{12}$$

We see clearly that only one dimensional imaging of the nucleon is relevant in inclusive DIS. We also see that the PDFs are defined in terms of QFT operators via the quark-quark correlator $\hat{\Phi}^{(0)}(x,p,S)$ by expending it in terms of γ-matrices and the corresponding basic Lorentz covariants.

To go to higher order of pQCD, we take the loop diagrams that describe gluon radiations and so on into account. After proper handling of these contributions, we obtain the factorized form[5] where the PDFs acquire the scale Q-dependence govern by QCD evolution equations. In practice, PDFs are parameterized and are given in the PDF library (PDFlib).

In summary, for studying one dimensional imaging of the nucleon with inclusive DIS, we took the following steps,

- General symmetry analysis leads to the general form of the hadronic tensor in terms of four independent structure functions.
- Parton model without QCD interaction leads to LO in pQCD and leading twist results in terms of Q-independent PDFs without (local) gauge invariance.
- Parton model with QCD multiple gluon scattering after collinear expansion leads to LO in pQCD, leading and higher twist contributions in terms of Q-independent but gauge invariant PDFs.
- Parton model with QCD multiple gluon scattering and "loop diagram contributions" after collinear approximation, regularization and renormalization leads to leading and higher order pQCD, leading twist contributions in factorized forms as functions of Q-evolved and gauge invariant PDFs.

In the following, I will follow this four steps and summarize what we achieved in the three dimensional case. Before that, I would like to recall the following two of historical developments that may be helpful for to us to construct the theoretical framework for the TMD case.

First, as mentioned, the study of three dimensional imaging of the nucleon was triggered by the experimental observation of SSA in the inclusive hadron-hadron collision. It was known that pQCD leads to negligibly small asymmetry for the hard part but the observed asymmetry can be as large as 40%. The hunting for such large asymmetries last for decades with following milestones:

- In 1991, Sivers introduced[6] the asymmetric quark distribution in a transversely polarized nucleon that is now known as the Sivers function.
- In 1993, Boros, Liang and Meng proposed[7] a phenomenological model that provides an intuitive physical picture showing that the asymmetry arises from the

orbital angular momenta of quarks and what they called "surface effect" caused by the initial or final state interactions.

- In 1993, Collins published[8] his proof that Sivers function has to vanish.
- In 2002, Brodsky, Hwang and Schmidt calculated[9] SSA for SIDIS using an explicit example where they took the orbital angular momentum of quark and the multiple gluon scattering into account.
- Collins pointed out[10] that the multiple gluon scattering is contained in the gauge link and that the conclusion of his proof in 1993 was incorrect because he forgot the gauge link; Belitsky, Ji and Yuan resolved[11, 12] the problem of defining the gauge link for a TMD parton density in light-cone gauge where the gauge potential does not vanish asymptotically.

Another historical development concerns the azimuthal asymmetry study in SIDIS. It was shown by Georgi and Politzer in 1977 that[13] final state gluon radiations lead to azimuthal asymmetries and could be used as a "clean test to pQCD". However, soon after, in 1978, it was shown by Cahn that[14] similar asymmetries can also be obtained if one includes intrinsic transverse momenta of partons. The latter, now named as Cahn effect, though power suppressed i.e. higher twist, can be quite significant and can not be neglected since the values of the asymmetries themselves are usually not very large.

The lessens that we learned from these histories are in particular the following two points, i.e., when studying TMDs,

- it is important to take the gauge link into account;
- higher twist effects can be important.

Both of them demand that, to describe SIDIS in terms of TMDs, we need the proper QFT formulation rather than the intuitive parton model.

3. TMDs Defined via Quark-Quark Correlator

The TMD PDFs of quarks are defined via the quark-quark correlator $\Phi^{(0)}$ given by Eq. (10) (after integration over k^-). A systematical study has been given in Ref.[15] and a very comprehensive treatment can also be found in Ref.[16]. Here, we first expand it in terms of γ-matrices and obtain a scalar, a pseudo scalar, a vector, an axial-vector and a tensor part. We then analyze the Lorentz structure of each part by expressing it in terms of possible "basic Lorentz covariants" and scalar functions. These scalar functions are known as TMD PDFs. There are totally 32 such TMD PDFs. Among them, 8 contribute at leading twist and they all have clear probability interpretations such as the number density, the helicity distribution, the transversity, the Sivers function, the Boer-Mulders function etc; 16 contribute at twist 3 and the other 8 contribute at twist 4. We emphasize that they are all scalar functions of x and $k_\perp$, i.e., depending on x and $k_\perp^2$. If we integrate over $d^2k_\perp$, terms where the basic Lorentz covariants are space odd vanish. At the leading twist, only 3 of 8 survive, i.e. the number density, the helicity distribution and the transversity.

Higher twist TMD PDFs are also defined via quark-j-gluon-quark correlators. Many of them are however not independent since they are related to those defined via the quark-quark correlator through the QCD equation of motion. It is interesting to see that,[25] although not generally proved, all the twist 3 TMD PDFs that are defined via quark-gluon-quark correlator and are involved in SIDIS are replaced by those defined via quark-quark correlator.

I also want to emphasize that fragmentation is just conjugate to parton distribution. We have one to one correspondence between TMD PDFs and TMD FFs.

4. Accessing the TMDs in High Energy Reactions

The TMDs can be studied in semi-inclusive high energy reactions such as SIDIS $e^- + N \rightarrow e^- + h + X$, semi-inclusive Drell-Yan $h + h \rightarrow l^+ + l^- + X$, and semi-inclusive hadron production in e^+e^--annihilation $e^+ + e^- \rightarrow h_1 + h_2 + X$. With SIDIS, we study TMD PDFs and TMD FFs, while with Drell-Yan and e^+e^- annihilation, we study TMD PDFs and TMD FFs separately. We now follow the same steps as those for inclusive DIS and briefly summarize what we already have in constructing the corresponding theoretical framework.

(I) The general forms of hadronic tensors: For all three classes of processes, the general forms of hadronic tensors have been studied and obtained. For SIDIS, it has been discussed in Refs.[17-20] and it has been shown that one need 18 independent structure functions for spinless h. For Drell-Yan, a comprehensive study was made in Ref. [21] and the number of independent structure functions is 48 for hadrons with spin 1/2. For e^+e^--annihilation, the study was presented in Ref.[22] and one needs 72 for spin-1/2 h_1 and h_2.

(II) LO in pQCD and leading twist parton model results: These are the simplest parton model results and can be obtained easily. E.g., for SIDIS, the result can be obtained from those given e.g. in Ref.[20] by neglecting all the power suppressed contributions. I emphasize that the result obtained this way is a complete parton model result at LO in pQCD and leading twist. It can be used to extract the TMDs at this order. Any attempt to go beyond LO in pQCD or to consider higher twists needs to go beyond this expression.

(III) LO in pQCD, leading and higher twist parton model results: For the semi-inclusive processes where only one hadron is involved, either in the initial or the final state, it has been shown[23-27] that the collinear expansion can be applied. Such processes include: semi-inclusive DIS $e^- + N \rightarrow e^- + q(jet) + X$, and e^+e^--annihilation $e^+ + e^- \rightarrow h + \bar{q}(jet) + X$. By applying the collinear expansion, we have constructed the theoretical frameworks for these processes with which leading as well as higher twist contributions can be calculated in a systematical way to LO in pQCD. The complete results up to twist-3 have been obtained in Refs.[25-27]. For unpolarized $e^- + N \rightarrow e^- + q(jet) + X$, the results up to twist 4 have also been obtained.[24] These results can be used as the basis for measuring these TMDs via the corresponding process at the LO in pQCD. I call in particular the attention to the results[27] for $e^+ + e^- \to h + \bar{q}(jet) + X$ for h with different spins. Those for spin-1 hadrons involve tensor polarization that is much less explored till now. See also talks by Y.K. Song and S.Y. Wei for more details.[28, 29]

However, for the above-mentioned three kinds of semi-inclusive processes, there are always two hadrons involved. Collinear expansion has not been proved how to apply for such processes. It is unclear how one can calculate leading and higher twist contributions in a systematical way. Nevertheless, twist 3 calculations that have been carried out for these processes,[30-32] practically in the following steps: (i) draw Feynman diagrams with multiple gluon scattering to the order of one gluon exchange, (ii) insert the gauge link in the correlator wherever needed to make it gauge invariant, (iii) carry out calculations to the order $1/Q$. Although not proved, it is interesting to see that the results obtained this way reduce exactly to those obtained in the corresponding simplified cases where collinear expansion is applied if we take the corresponding fragmentation functions as δ-functions.

(IV) TMD factorization and evolution: TMD factorization theorem has been established at the leading twist for semi-inclusive processes[33-39] TMD evolution theory is also developing very fast[40-50] See the overview talk by Daniel Boer.[51]

5. TMD Parameterizations

Experiments have been carried out for all three kinds of semi-inclusive reactions. The results are summarized in particular in the plenary talks in this conference by Marcin Stolarski, Renee Fatemi, and Armine Rostomyan. Here, I will just try to sort out the TMD parameterizations that we already have.

The first part concerns what people called "the first phase parameterizations", i.e. TMD parameterizations without QCD evolutions.

(1) Transverse momentum dependence: This is usually taken as[60-56] a Gaussian in a factorized form independent of the longitudinal variable z or x. The width has been fitted, the form and flavor dependence etc have been tested. Roughly speaking, this is a quite satisfactory fit. However, it has also been pointed out, e.g. in [55] for the TMD FF, that the Gaussian form seems to depend on the flavor and even on z, which means that it is only a zeroth order approximation.

(2) Sivers function: It is usually parameterized[57-63] in the form of the number density $f_q(x, k_\perp)$ multiplied by a x-dependent factor $\mathcal{N}_q(x)$ and a $k_\perp$-dependent factor $h(k_\perp)$, and $\mathcal{N}_q(x) \sim x^\alpha (1-x)^\beta$ while $h(k_\perp)$ is taken as a Gaussian. There exist already different sets such as the Bochum, the Torino and the Vogelsang-Yuan fits. One thing seems to be clear that the Sivers function is nonzero for proton and it has different signs for u- and d-quark.

(3) Transversity and Collins function: A simultaneous extraction of them from SIDIS data have been carried out by the Torino group.[53, 64] A similar form as that for the Sivers function has been taken and it has been obtained that also the Collins function is nonzero and has different signs e.g. for $u \to \pi^+$ or $d \to \pi^+$.

(4) Boer-Mulders function: Clear signature for non-zero Boer-Mulders function has been obtained from SIDIS data on $\langle \cos 2\phi \rangle$ asymmetry[65]−.[69] The form was taken again similar to the Sivers function. However, I would like to point out that the $\langle \cos 2\phi \rangle$ asymmetry receives twist-4 contributions due to the Cahn effect.[14] A proper treatment of such twist-4 effect involves twist-4 TMDs as discussed in Ref. [24]. Because of the multiple gluon scattering shown in Fig. 1, the twist-4 effects could be very much different from that given in [14].

Attempts to parameterize other TMDs such as pretzelocity $h_{1T}^{\perp}$ have also been made.[70] Although there is no enough data to give high accuracy constraints, the qualitative features obtained are also interesting.

The second part concerns the TMD evolution. As mentioned earlier, this is a topic that develops very fast recently. A partial list of recent dedicated publications is Refs. [40-50]. QCD evolution equations have been constructed. The numerical results obtained show clearly that TMD evolution is quite significant and it is important to use the comprehensive TMD evolution rather than a separate evolution of the transverse and longitudinal dependences respectively. See the overview talk by Daniel Boer[51] at this conference.

At last, I want to mention the first version of TMD PDFlib has already created.[71]

6. Summary and Outlook

In summary, I just want to emphasize that three dimensional imaging of the nucleon is a hot and fast developing topic in last years. Many progresses have been made and many questions are open. Especially in view of the running and planned facilities such as the electron-ion colliders, we expect even rapid development in next years.

I apologize for many aspects that I could not cover such as TMDs and Wigner function, model calculations, nuclear dependences, and hyperon polarization. The readers are referred to many interesting talks at this conference.

I thank X.N. Wang, Y.K. Song, S.Y. Wei, K.B. Chen, J.H. Gao and many other people for collaboration and help in preparing this talk. My sincere thanks also go to John Collins for communications. This work was supported in part by NSFC (Nos.11035003 and 11375104), and the Major State Basic Research Development Program in China (No. 2014CB845406).

References

1. R.P. Feynman, *Photon Hadron Interactions*, W.A. Benjamin. 1972.
2. R. K. Ellis, W. Furmanski and R. Petronzio, Nucl. Phys. B **207**, 1 (1982).
3. R. K. Ellis, W. Furmanski and R. Petronzio, Nucl. Phys. B **212**, 29 (1983).
4. J. -w. Qiu and G. F. Sterman, Nucl. Phys. B **353**, 105 (1991); B **353**, 137 (1991).
5. J. C. Collins *et al.*, Adv. Ser. Direct. High Energy Phys. **5**, 1 (1988).
6. D. W. Sivers, Phys. Rev. D **41**, 83 (1990); **43**, 261 (1991).
7. C. Boros, Z. T. Liang and T. C. Meng, Phys. Rev. Lett. **70**, 1751 (1993).
8. J. C. Collins, Nucl. Phys. B **396**, 161 (1993).
9. S. J. Brodsky, D. S. Hwang and I. Schmidt, Phys. Lett. B **530**, 99 (2002).

10. J. C. Collins, Phys. Lett. B **536**, 43 (2002).
11. X. d. Ji and F. Yuan, Phys. Lett. B **543**, 66 (2002).
12. A. V. Belitsky, X. Ji and F. Yuan, Nucl. Phys. B **656**, 165 (2003).
13. H. Georgi and H. Politzer, Phys. Rev. Lett. **40**, 3 (1978).
14. R. N. Cahn, Phys. Lett. B **78**, 269 (1978).
15. K. Goeke, A. Metz and M. Schlegel, Phys. Lett. B **618**, 90 (2005).
16. P. Mulders, talk in this conference, and lectures in 17th Taiwan nuclear physics summer school, Aug. 25-28, 2014.
17. M. Gourdin, Nucl. Phys. B **49**, 501 (1972).
18. A. Kotzinian, Nucl. Phys. B **441**, 234 (1995).
19. M. Diehl and S. Sapeta, Eur. Phys. J. C **41**, 515 (2005).
20. A. Bacchetta *et al.*, JHEP **0702**, 093 (2007).
21. S. Arnold, A. Metz and M. Schlegel, Phys. Rev. D **79**, 034005 (2009).
22. D. Pitonyak, M. Schlegel and A. Metz, Phys. Rev. D **89**, no. 5, 054032 (2014).
23. Z. -t. Liang and X. -N. Wang, Phys. Rev. D **75**, 094002 (2007).
24. Y. -k. Song, J. -h. Gao, Z. -t. Liang and X. -N. Wang, Phys. Rev. D **83**, 054010 (2011).
25. Y. -k. Song, J. -h. Gao, Z. -t. Liang and X. -N. Wang, Phys. Rev. D **89**, 014005 (2014).
26. S. -y. Wei, Y. -k. Song and Z. -t. Liang, Phys. Rev. D **89**, 014024 (2014).
27. S. Y. Wei, K. b. Chen, Y. k. Song and Z. t. Liang, arXiv:1410.4314 [hep-ph].
28. Y.K. Song, talk given at this conference.
29. S.Y. Wei, talk given at this conference.
30. P. J. Mulders and R. D. Tangerman, Nucl. Phys. B **461**, 197 (1996).
31. D. Boer, R. Jakob and P. J. Mulders, Nucl. Phys. B **504**, 345 (1997).
32. Z. Lu and I. Schmidt, Phys. Rev. D **84**, 114004 (2011).
33. J. C. Collins and D. E. Soper, Nucl. Phys. B **193**, 381 (1981) [Erratum-ibid. B **213**, 545 (1983)]; Nucl. Phys. B **194**, 445 (1982).
34. J. C. Collins, D. E. Soper and G. F. Sterman, Nucl. Phys. B **250**, 199 (1985).
35. J. C. Collins, D. E. Soper and G. F. Sterman, Nucl. Phys. B **261**, 104 (1985).
36. X. d. Ji, J. P. Ma and F. Yuan, Phys. Lett. B **610**, 247 (2005).
37. A. Idilbi, X. d. Ji, J. P. Ma and F. Yuan, Phys. Rev. D **70**, 074021 (2004).
38. X. d. Ji, J. P. Ma and F. Yuan, Phys. Lett. B **597**, 299 (2004).
39. X. d. Ji, J. p. Ma and F. Yuan, Phys. Rev. D **71**, 034005 (2005).
40. A. A. Henneman, D. Boer and P. J. Mulders, Nucl. Phys. B **620**, 331 (2002).
41. J. Zhou, F. Yuan and Z. T. Liang, Phys. Rev. D **79**, 114022 (2009).
42. Z. B. Kang, B. W. Xiao and F. Yuan, Phys. Rev. Lett. **107**, 152002 (2011).
43. S. M. Aybat and T. C. Rogers, Phys. Rev. D **83**, 114042 (2011).
44. S. M. Aybat *et al.*, Phys. Rev. D **85**, 034043 (2012).
45. M. Anselmino, M. Boglione and S. Melis, Phys. Rev. D **86**, 014028 (2012).
46. P. Sun and F. Yuan, Phys. Rev. D **88**, no. 11, 114012 (2013).
47. M. G. Echevarria *et al.*, Phys. Rev. D **89**, 074013 (2014).
48. C. A. Aidala *et al.*, Phys. Rev. D **89**, 094002 (2014).
49. Z. B. Kang, A. Prokudin, P. Sun and F. Yuan, arXiv:1410.4877 [hep-ph].
50. M. G. Echevarria, A. Idilbi and I. Scimemi, Phys. Rev. D **90**, 014003 (2014).
51. D. Boer, talk given at this conference.
52. M. Anselmino *et al.*, Phys. Rev. D **71**, 074006 (2005).
53. M. Anselmino *et al.*, Phys. Rev. D **75**, 054032 (2007).
54. P. Schweitzer, T. Teckentrup and A. Metz, Phys. Rev. D **81**, 094019 (2010).
55. A. Signori, A. Bacchetta, M. Radici and G. Schnell, JHEP **1311**, 194 (2013).
56. M. Anselmino *et al.*, JHEP **1404**, 005 (2014).
57. A. V. Efremov *et al.*, Phys. Lett. B **612**, 233 (2005).

58. J. C. Collins *et al.*, Phys. Rev. D **73**, 014021 (2006).
59. S. Arnold *et al.*, arXiv:0805.2137 [hep-ph].
60. M. Anselmino *et al.*, Phys. Rev. D **71**, 074006 (2005).
61. M. Anselmino *et al.*, Eur. Phys. J. A **39**, 89 (2009).
62. W. Vogelsang and F. Yuan, Phys. Rev. D **72**, 054028 (2005).
63. A. Bacchetta and M. Radici, Phys. Rev. Lett. **107**, 212001 (2011).
64. M. Anselmino *et al.*, Phys. Rev. D **87**, 094019 (2013).
65. B. Zhang, Z. Lu, B. Q. Ma and I. Schmidt, Phys. Rev. D **77**, 054011 (2008).
66. B. Zhang, Z. Lu, B. Q. Ma and I. Schmidt, Phys. Rev. D **78**, 034035 (2008).
67. V. Barone, A. Prokudin and B. Q. Ma, Phys. Rev. D **78**, 045022 (2008).
68. V. Barone, S. Melis and A. Prokudin, Phys. Rev. D **81**, 114026 (2010).
69. Z. Lu and I. Schmidt, Phys. Rev. D **81**, 034023 (2010).
70. J. Zhu and B. Q. Ma, Phys. Rev. D **82**, 114022 (2010).
71. F. Hautmann *et al.*, Eur. Phys. J. C **74**, no. 12, 3220 (2014).

Spin Physics (SPIN2014)
International Journal of Modern Physics: Conference Series
Vol. 40 (2016) 1660009 (11 pages)
© The Author(s)
DOI: 10.1142/S2010194516600090

Spin Physics at J-PARC

S. Kumano

KEK Theory Center, Institute of Particle and Nuclear Studies, KEK
1-1, Ooho, Tsukuba, Ibaraki, 305-0801, Japan
J-PARC Branch, KEK Theory Center, Institute of Particle and Nuclear Studies, KEK
and Theory Group, Particle and Nuclear Physics Division, J-PARC Center
203-1, Shirakata, Tokai, Ibaraki, 319-1106, Japan
shunzo.kumano@kek.jp

Published 29 February 2016

Spin-physics projects at J-PARC are explained by including future possibilities. J-PARC is the most-intense hadron-beam facility in the high-energy region above multi-GeV, and spin physics will be investigated by using secondary beams of kaons, pions, neutrinos, muons, and antiproton as well as the primary-beam proton. In particle physics, spin topics are on muon $g-2$, muon and neutron electric dipole moments, and time-reversal violation experiment in a kaon decay. Here, we focus more on hadron-spin physics as for future projects. For example, generalized parton distributions (GPDs) could be investigated by using pion and proton beams, whereas they are studied by the virtual Compton scattering at lepton facilities. The GPDs are key quantities for determining the three-dimensional picture of hadrons and for finding the origin of the nucleon spin including partonic orbital-angular-momentum contributions. In addition, polarized parton distributions and various hadron spin topics should be possible by using the high-momentum beamline. The strangeness contribution to the nucleon spin could be also investigated in principle with the neutrino beam with a near detector facility.

Keywords: J-PARC; nucleon; spin; QCD; $g-2$; electric dipole moment; kaon decay.

PACS numbers: 13.85.−t, 24.85.+p, 12.38.−t, 13.40.Em, 13.20.Eb

1. Introduction to J-PARC

Japan Proton Accelerator Research Complex (J-PARC) is located at Tokai in Japan,[1] and it is a joint facility between KEK (High Energy Accelerator Research Organization) and JAEA (Japan Atomic Energy Agency) for projects in wide fields of science. KEK is in charge of the particle- and nuclear-physics projects at the 50-GeV proton synchrotron. J-PARC provides most intense proton beam in the energy region above multi-GeV. Nuclear and particle physics projects use secondary beams such as kaons, pions, neutrinos, muons, and antiproton as well as the primary 50-GeV proton beam. The J-PARC experiments have been started for neutrino oscillations and strangeness hadron experiments. In this report, we explain spin physics at J-PARC, mainly on possibilities of hadron spin physics.

J-PARC could cover a wide range of spin projects from fundamental particle physics to hadron spin physics. They include muon $g-2$, muon and neutron electric dipole moments, and time-reversal violation experiment in a kaon decay as particle-physics projects. In hadron physics, there are possible projects for clarifying the origin of nucleon spin and associated three-dimensional structure of the nucleon.

J-PARC facility

A bird's-eye view of the J-PARC facility is shown in Fig. 1. The accelerator consists of a linac as an injector, a 3-GeV rapid cycling synchrotron, and a 50-GeV synchrotron. The energy of the 50-GeV synchrotron is 30 GeV at this stage. J-PARC provides most intense proton beam in the high-energy region ($E > 1$ GeV), and it is 1 MW in the 3-GeV synchrotron and 0.75 MW is expected in the 50-GeV one. There are three major projects at J-PARC:

- material and life sciences as well as particle physics with neutrons and muons produced by the 3-GeV proton beam,
- nuclear and particle physics with secondary beams (pions, kaons, neutrinos, muons, and antiprotons) by the 50-GeV proton beam and also with protons of the 50-GeV primary beam,
- nuclear transmutation and neutron physics by the linac.

Hadron-physics projects are investigated at the hadron experimental facility in Fig. 1. Experiments on lepton-flavor violation and time-reversal violation experiment in a kaon decay will be done also in this hadron hall, in which the beam-layout plan is shown in Fig. 2. The K1.8 is intended to have kaons with momentum around 1.8 GeV/c for the studies, for example, on strangeness -2 hypernuclei with Ξ^- by (K^-, K^+) reactions. The K1.1/0.8 beamline is designed for low-momentum stopped kaon experiments such as the studies of kaonic nuclei. The neutral kaon beamline

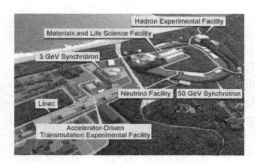

Fig. 1.   J-PARC facility.[1]

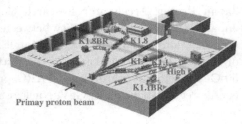

Fig. 2.   Beamline layout of hadron hall.[1]

(KL) is for studying CP violating processes such as $K_L \to \pi^0 \nu \bar{\nu}$. The "High p" in Fig. 2 indicates the high-momentum beamline for 50-GeV protons and unseparated hadrons. In the beginning stage of J-PARC, the proton-beam energy is 30 GeV instead of the original plan of 50 GeV. Measurements on muon $g-2$, muon and neutron electric dipole moment will be done at the Materials and Life Science Experimental Facility (MLF) and at the linac part for neutron physics. Furthermore, hadron spin physics could be possible in principle with the neutrino beam with a near detector, for example, by focusing on strangeness spin in the axial form factor.

In future, hadron spin projects will become possible at the hadron hall because the high-momentum beamline in Fig. 2 will be ready in the Japanese fiscal year of 2016. The high-momentum beamline can transport the primary proton beam (30 GeV) with the intensity of 10^{10}-10^{12}/sec to the Hadron Hall, and it is a branch from the main proton beam of 10^{13} or 10^{14}/sec. In addition, we can obtain unseparated secondary beams such as pions etc. The beam intensity of these secondaries depends on its species and momentum. A typical intensity for 10 GeV/c (15 GeV/c) pions would be in the order of 10^7/sec (10^6/sec). Because there is no approved proposal to study hadron spin physics at J-PARC at this stage, we need to develop such possibilities. In this article, we first introduce particle-spin projects, and then we focus our discussions on high-energy hadron projects.

2. Spin in Particle Physics

Particle-spin projects are intended to probe physics beyond the standard model. Here, we introduce three projects on (1) muon $g-2$ and electric dipole moment, (2) muon polarization in kaon decay, and (3) electric dipole moment of the neutron.

2.1. *Muon $g-2$ and electric dipole moment*

There is a project to measure the anomalous magnetic moment $(g-2)$ and electric dipole moment (EDM) of the muon at the MLF of J-PARC.[2] There is a long history of measurements on the muon $g-2$. The muon magnetic moment is given by its spin and gyromagnetic ratio g_μ as $\vec{\mu}_\mu = g_\mu \frac{e}{2m} \vec{s}$. The anomalous magnetic moment a_μ is related to $g_\mu - 2$ as $a_\mu = (g_\mu - 2)/2$. We denote the muon EDM as d_μ. The electric dipole moment is given by the spin and η_μ as $\vec{d}_\mu = \eta_\mu \frac{e\hbar}{2m} \vec{s}$. The most recent measurements were done by the BNL-E821 experiment, and a_μ was measured down to 0.54 ppm and d_μ to 1.9×10^{-19}e·cm.[2] It is especially interesting that the current value $a_\mu = 0.00116592089(63)$ deviates from the theoretical value $0.00116591828(49)$ with 3.3σ discrepancy. This deviation exists even if theoretical uncertainties, mainly from hadronic corrections, are taken into account, so that it could suggest new physics. Therefore, it is important to improve the experimental measurement by an independent method.

Under the static electric and magnetic fields, the muon spin precesses with the frequency

$$\vec{\omega} = -\frac{e}{m}\left[a_\mu\vec{B} - \left(a_\mu - \frac{1}{\gamma^2 - 1}\right)\frac{\vec{\beta}\times\vec{E}}{c} - \frac{e}{m}\left[\frac{\eta}{2}\left(\vec{\beta}\times\vec{B} + \frac{\vec{E}}{c}\right)\right] \equiv \vec{\omega}_a + \vec{\omega}_\eta, \quad (1)$$

where $\vec{\omega}_a$ is the precession vector due to the anomalous magnetic moment, and $\vec{\omega}_\eta$ is the one due to the electric dipole moment. In the CERN and BNL experiments, the muon energy was chosen to terminate the contribution of the $\vec{\beta}\times\vec{E}$ term, and the EDM contribution $\vec{\omega}_\eta$ is neglected. In the proposed J-PARC experiment in Fig. 3, the electric field $\vec{E}$ is terminated to give the precession frequency

$$\vec{\omega} = -\frac{e}{m}\left[a_\mu\vec{B} + \frac{\eta}{2}\vec{\beta}\times\vec{B}\right]. \quad (2)$$

Here, the first $g-2$ term and the second EDM term are orthogonal with each other, and they can be separated by appropriate experimental design. The purpose of this experiment is to improve the current limit of $g-2$ to 0.1 ppm by a factor of five and the EDM to 10^{-21} e·cm by a factor of two orders of magnitude than the current limit. R&D is in progress for significant technological developments to start this experiment within 2010's.[2]

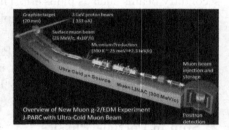

Fig. 3. $g-2$ and EDM experiment.[2]

2.2. Transverse muon polarization in kaon decay

There is a project to investigate time-reversal violation by the transverse muon polarization $P_T = \hat{s}_\mu \cdot (\hat{p}_{\mu^+} \times \hat{p}_{\pi^0})$ in kaon decay $K^+ \to \pi^0\mu^+\nu$ at J-PARC.[3] It is called TREK (Time-Reversal violation Experiment with Kaons) experiment. This polarization could probe a signature beyond the standard model because the standard-model contributions from higher-order effects are considered to less than 10^{-6}. The most recent measurement was done by the KEK-E246 experiment and it was $P_T = -0.0017 \pm 0.0023\,(\text{stat}) \pm 0.0017\,(\text{syst})$, which is consistent with no T-violation. New-physics models suggest that the polarization value should be as large as 10^{-3}, which is just below the KEK-E246 limit. Therefore, it is valuable to measure the polarization more accurately to provide a clue for new physics. In the proposed J-PARC experiment, background reduction will be done by improved detector, especially the upgrade of polarimeter and magnet. The resulting error is expected to be $\delta P_T = 10^{-4}$. In the beginning low-intensity period of J-PARC, the collaboration proposed to start another experiment E36 on the decay ratio $R_K = \Gamma(K^+ \to e^+\nu)/\Gamma(K^+ \to \mu^+\nu)$ to test lepton universality by using a subsystem of the TREK experiment.[3]

2.3. *Electric dipole moment of the neutron*

Ultracold neutrons (UCN) are used for measurements of fundamental physics quantities, and new generation UCN sources are under developments for experiments at facilities including J-PARC.[4] In particular, the neutron electric dipole moment (nEDM) should be able to probe new physics. For example, the supersymmetric model indicates that the nEDM is of the order of 10^{-27} e·cm, which is just below the limit 3×10^{-26} e·cm obtained by a Grenoble experiment. Since the limit is constrained by the statistical error, efforts have been made to increase the UCN density, as well as efforts to reduce the systematic error. Spallation UCN sources are considered for the nEDM measurement at J-PARC. Due to the increase of the proton-beam power to 20 kW, the density is expected to become 10^3-10^4 UCN/cm^3, whereas it was 0.7 UCN/cm^3 in the Grenoble experiment. With the new UCN sources together with the intense proton beam, much improvement will be made for the upper bound of the nEDM into the region of 10^{-28} e·cm. At J-PARC, the nEDM experiment is considered at the linac section.[4]

3. Spin in Hadron Physics

3.1. *Flavor dependence of antiquark distributions*

There are proposals on dimuon experiments by using primary proton beam for measuring parton distribution functions (PDFs) in the medium Bjorken-x region.[5] A typical example is shown in Fig. 4 for measuring the light-antiquark distribution ratio $\bar{d}(x)/\bar{u}(x)$ by the Drell-Yan processes of pp and pd, where p (d) is proton (deuteron). According to perturbative QCD, the $\bar{u}/\bar{d}$ asymmetry should be very small. Experimentally, the difference was sug-

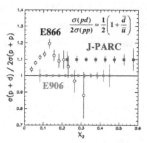

Fig. 4. Drell-Yan cross-section ratios.[5]

gested by the violation of the Gottfried sum rule,[6] and it was confirmed explicitly by the Fermilab-E866 measurements in Fig. 4. However, there is a peculiar tendency of $\bar{d}/\bar{u} < 1$ as x becomes larger, which is difficult to be understood theoretically. The E906/SeaQuest experiment at Fermilab is in progress for dimuon experiments to test this tendency, and it could be continued at J-PARC.

Measurements of antiquark distributions in nuclei are also interesting. The Fermilab Drell-Yan measurements showed that nuclear modifications of the antiquark distributions are very small at $x \sim 0.1$, which is in contradiction to the conventional pion-excess contribution. It is important to confirm this result by an independent experiment and to extend the measured x region for finding a physics mechanism of nuclear medium effects on the antiquark distributions. In addition, determination of parton distribution functions (PDFs) at large x is valuable for precisely calculating other high-energy reactions, for example, high-p_T jet and hadron production cross sections at LHC.

3.2. *Generalized parton distributions*

Generalized parton distributions (GPDs) are key quantities for studying three-dimensional structure of hadrons, and hence to clarify the origin of the nucleon spin including partonic orbital-angular-momentum contributions.[7] They are measured typically in deeply virtual Compton scattering (DVCS) as shown in Fig. 5. However, there are possibilities to investigate them at hadron facilities like J-PARC.

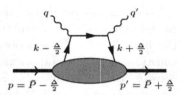

Fig. 5.   Kinematics for GPDs.

First, we define kinematical variables for the process $\gamma^* + p \rightarrow \gamma + p$ by the momenta given in Fig. 5. The average nucleon and photon momenta ($\bar{P}$ and $\bar{q}$) and the momentum transfer Δ are defined by $\bar{P} = (p+p')/2$, $\bar{q} = (q+q')/2$, $\Delta = p'-p = q-q'$. Then, Q^2 and t are given by $Q^2 = -q^2$, $\bar{Q}^2 = -\bar{q}^2$, $t = \Delta^2$, and the generalized scaling variable x and a skewdness parameter ξ are defined by $x = Q^2/(2p \cdot q)$, $\xi = \bar{Q}^2/(2\bar{P} \cdot \bar{q})$. The variable x indicates the lightcone momentum fraction carried by a quark in the nucleon. The skewdness parameter ξ or the momentum Δ indicates the momentum transfer from the initial nucleon to the final one or the one between the quarks. The GPDs for the nucleon are given by off-forward matrix elements of quark and gluon operators with a lightcone separation between nucleonic states. The quark GPDs are defined by

$$
\int \frac{dy^-}{4\pi} e^{ix\bar{P}^+ y^-} \left\langle p' \left| \overline{\psi}(-y/2)\gamma^+\psi(y/2) \right| p \right\rangle \Big|_{y^+=\vec{y}_\perp=0}
$$
$$
= \frac{1}{2\bar{P}^+} \, \overline{u}(p') \left[H_q(x,\xi,t)\gamma^+ + E_q(x,\xi,t)\frac{i\sigma^{+\alpha}\Delta_\alpha}{2\,M} \right] u(p), \tag{3}
$$

where $H_q(x,\xi,t)$ and $E_q(x,\xi,t)$ are the unpolarized GPDs for the nucleon.

The major properties of the GPDs are the following. In the forward limit (Δ, ξ, $t \rightarrow 0$), the nucleonic GPDs $H_q(x,\xi,t)$ become usual PDFs: $H_q(x,0,0) = q(x)$. Next, their first moments are the Dirac and Pauli form factors of the nucleon: $\int_{-1}^{1} dx H_q(x,\xi,t) = F_1(t)$, $\int_{-1}^{1} dx E_q(x,\xi,t) = F_2(t)$. Furthermore, the second moment is related to the quark orbital-angular-momentum contribution (L_q) to the nucleon spin: $J_q = \frac{1}{2} \int dx\, x\, [H_q(x,\xi,t=0) + E_q(x,\xi,t=0)] = \frac{1}{2}\Delta q + L_q$. From this relation, we expect that the origin of nucleon spin will be clarified by including the orbital-angular-momentum contributions. The GPDs contain information on the longitudinal momentum distributions and transverse structure as the form factors, so that they are appropriate quantities for understanding the three-dimensional structure of hadrons. The GPDs can be measured by the virtual Compton scattering in Fig. 5 at lepton facilities. However, they can be measured also at hadron facilities such as J-PARC and GSI-FAIR, and examples are explained in the following subsections.

3.2.1. *Exclusive Drell-Yan for studying GPDs*

The unseparated hadron beam, which is essentially the pion beam, will become available at the high-momentum beamline of J-PARC, so the exclusive Drell-Yan process is a possible future project at J-PARC. In Fig. 6, the exclusive dimuon process $\pi^- + p \to \mu^+\mu^- + B$ is shown.[8] The process probes the nucleonic GPDs if B is the nucleon, whereas it is related to the transi-

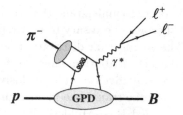

Fig. 6. Exclusive Drell-Yan.

tion GPDs for $B \neq N$. The cross section contains not only the GPDs but also the pion distribution amplitude. Although there are models for this quantity such as the asymptotic form or the Chernyak-Zhitnitsky type, the pion part is rather well studied and tested experimentally by the $\gamma\pi$ transition of Belle and BaBar. Using such studies for constraining the pion distribution amplitude, we should be able to obtain information on the GPDs by the exclusive Drell-Yan process.

3.2.2. *GPDs in the ERBL region*

Using the high-energy proton beam, we could investigate the GPDs at hadron facilities by using exclusive hadron-production reactions $a + b \to c + d + e$ such as $N + N \to N + \pi + B$, as shown in Fig. 7.[9] The GPD has three kinematical regions, (1) $-1 < x < -\xi$, (2) $-\xi < x < \xi$, (3) $\xi < x < 1$, as shown in Fig. 8. The intermediate region (2) indicates an emission of quark with momentum fraction $x + \xi$ with an emission of antiquark with momentum fraction $\xi - x$. The regions (1) and (3) are called as DGLAP (Dokshitzer-Gribov-Lipatov-Altarelli-Parisi) regions, and (2) is called the ERBL (Efremov-Radyushkin-Brodsky-Lepage) region. If the hadrons c and d have large and nearly opposite transverse momenta and a large invariant energy, so that an intermediate exchange could be considered as a $q\bar{q}$ state. Then, the $q\bar{q}$ attached to the nucleon is expressed by the GPDs in a special kinematical region of the ERBL which is in the middle of three regions of Fig. 8. This method of measuring the GPDs is complementary to the virtual Compton scattering in the sense that the specific ERBL region is investigated in the medium x region due to the J-PARC energy of 30 GeV.

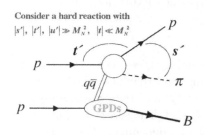

Fig. 7. GPD studies in $2 \to 3$ process.[9]

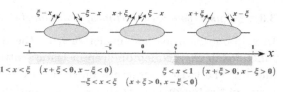

Fig. 8. Three kinematical regions of GPDs.

3.3. *Transverse-momentum-dependent distributions*

In order to understand the origin of nucleon spin including orbital angular momenta, it becomes necessary to understand the three-dimensional structure of the nucleon. The momentum distribution along the motion of the nucleon is the longitudinal parton distribution function, and the transverse distribution plays a role of form factor at given x. The inclusive Drell-Yan ($p + p \rightarrow \mu^+\mu^- + X$) measurements can probe transverse-momentum-dependent (TMD) polarized parton distributions, so called Boer-Mulders (BM) functions, by observing violation of the Lam-Tung relation in the cross sections.[5] The BM functions indicate transversely-polarized quark distributions in the unpolarized nucleon.

Other interesting TMD distributions are the Sivers functions, which indicate unpolarized quark distributions in the transversely-polarized nucleon. For example, they appear in single spin asymmetries of hadron-production processes $p + \vec{p} \rightarrow h+X$. Their measurements are valuable not only for understanding transverse structure but also for finding a relation to semi-inclusive lepton measurement because the TMD distributions change sign due to a difference of the gauge link. It is analogous to the Aharonov-Bohm effect.[10] In order to show the advantage of J-PARC measurements, we show single spin asymmetries of D-meson production at J-PARC and RHIC by considering the Sivers' mechanism in Fig. 9.[5,11] The figures indicate the advantage that the quark (gluon) Sivers functions are determined well at J-PARC (RHIC).

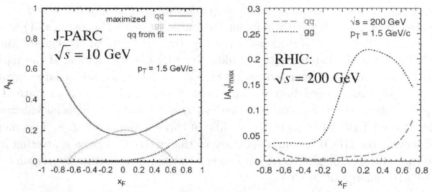

Fig. 9.   Single spin asymmetry in D-meson production at J-PARC and RHIC.[5,11]

3.3.1. *Tensor-polarized parton distribution functions*

In spin-one hadrons and nuclei such as the deuteron, there exist new structure functions in charged-lepton deep inelastic scattering, and they are called b_1, b_2, b_3, and b_4. The twist-2 structure functions are b_1 and b_2, and they are expressed in terms of the tensor-polarized quark distributions $\delta_T q(x)$. The b_1 was first measured by the HERMES collaboration; however, the detail of x dependence is not clear and there is no accurate data at medium and large x. The JLab b_1 proposal was approved and its measurement will start in a few years. The HERMES data indicated the violation of the sum rule $\int dx b_1(x) = 0$,[12] which suggests the existence of finite tensor polarization in antiquark distributions. This finite $\delta_T \bar{q}(x)$ can be directly measured at hadron facilities by Drell-Yan processes with tensor-polarized deuteron ($p +$ $\vec{d} \rightarrow \mu^+ \mu^- + X$).[12] We note that the polarized-proton beam is not

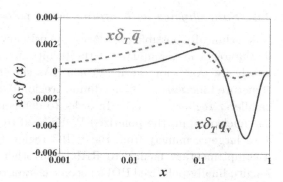

Fig. 10.   Tensor-polarized PDFs.[12]

needed for this experiment. This experiment is complementary to the HERMES and JLab measurements of b_1 in the sense that the antiquark tensor polarization can be measured specifically. The experiment will clarify an exotic dynamical aspect of the deuteron in terms of quark and gluon degrees of freedom, which cannot be done in low-energy measurements.

3.3.2. *Elastic single-spin asymmetry*

The origin of the nucleon spin is one of unsolved issues in hadron physics; however, there is another mysterious experimental result in elastic spin symmetries. For example, measurements indicated that the single-spin asymmetry in $p\bar{p}$ increases as $p_\perp$ becomes larger at AGS in Fig. 11.[13] According to perturbative QCD, the asymmetry has to vanish at $p_\perp \rightarrow \infty$. Because the data are taken up to $p_\perp^2 = 8$ GeV^2, nonperturbative physics might have contributed to the finite asymmetry. Considering the peculiar feature of the AGS data, we need to confirm the experimental measurements by an independent facility such as J-PARC before discussing possible physics mechanism.

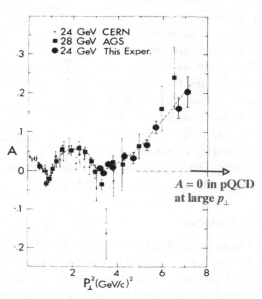

Fig. 11.   Elastic single spin asymmetry in $p\bar{p}$.[13]

However, rather than a mere confirmation, innovative methods should be developed, such as by changing targets and by considering angular distributions in order to provide clear evidences for theorists to understand the mechanism.

3.4. *Spin physics with proton-beam polarization*

It is technically possible to polarize the primary proton beam at J-PARC.[5] However, it requires a major update of the facility, we need to propose significant projects which are worth for the investment. As shown in Fig. 4, J-PARC could measure structure functions in the medium-x region $0.2 < x < 0.7$, whereas RHIC probes a smaller-x region ($x \sim 0.1$). In order to determine all the partonic contributions to the nucleon spin, the polarized PDFs need to be precisely understood from small x to large x, namely from the RHIC region to the J-PARC one. Therefore, it is a complementary facility to RHIC and other high-energy accelerators. Although longitudinally-polarized PDFs become clearer recently by various hadron and lepton reactions as shown in Fig. 12, their flavor decomposition and gluon distribution are not obvious yet. If the proton beam is polarized with the designed energy 50 GeV, the J-PARC facility could significantly contribute to the clarification of the origin of the nucleon spin.

In addition, there is new transverse spin physics, such as twist-two transversity distributions, by measuring double spin asymmetries of Drell-Yan processes with transversely polarized protons. The quark transversity distributions are unique in the sense that they do not couple to the gluon polarization due to their chiral-odd property. They are very different from the longitudinally-polarized PDFs. The transversity distributions can be measured at medium x with the polarized proton beam at J-PARC.

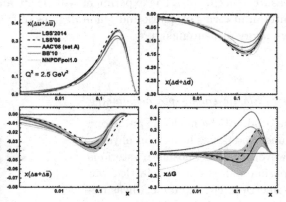

Fig. 12.　Situation of polarized PDFs.[14]

4. Summary

A wide range of spin projects are possible in particle and nuclear physics at J-PARC. As the particle-physics topics, we introduced muon $g-2$, muon and neutron electric dipole moments, and time-reversal violation experiment in a kaon decay. They could probe physics beyond the standard model by precision measurements with the high-intensity advantage of the J-PARC facility, and the measurements should shed light on new physics direction. In hadron spin physics, the studies of GPDs and TMDs should clarify the origin of the nucleon spin and also the three-dimensional structure of the nucleon including the transverse structure. In particular, the high-momentum beamline will be ready soon, and it can be used for various topics on high-energy hadron spin physics.

Acknowledgments

This work was partially supported by the MEXT KAKENHI Grant Number 25105010.

References

1. http://j-parc.jp/index-e.html. S. Kumano, Nucl. Phys. A 782 (2007) 442; AIP Conf. Proc. 1056 (2008) 444; J. Phys. Conf. Ser. 312 (2011) 032005.
2. J-PARC proposal P34, M. Aoki *et al.* (2009); J-PARC E34 collaboration, Conceptual Design Report (2011); N. Saito, AIP Conf.Proc. 1467 (2012) 45.
3. J-PARC proposal P06, K. Paton *et al.* (2006); P36, C. Rangacharyulu *et al.* (2010).
4. J-PARC UCN Taskforce, arXiv:0907.0515; J-PARC proposal P33, Y. Arimoto *et al.* (2009); Y. Masuda *et al.*, JPS Conference Proceedings of the 2nd International Symposium on Science at J-PARC.
5. J-PARC proposal P04, J. Chiba *et al.* (2006); proposal P24, M. Bai *et al.* (2008).
6. S. Kumano, Phys. Rept. 303 (1998) 183; G.T. Garvey and J.-C. Peng, Prog. Part. Nucl. Phys. 47 (2001) 203; J.-C. Peng, J.-W. Qiu, Prog. Part. Nucl. Phys. 76 (2014) 43.
7. M. Diehl, Phys. Rept. 388 (2003) 41; M. Diehl and P. Kroll, Eur. Phys. J. C 73 (2013) 2397.
8. E. R. Berger, M. Diehl, and B. Pire, Phys. Lett. B 523 (2001) 265; W.-C. Chang *et al.*, in preparation. Presentations by W.-C. Chang, P. Kroll, K. Tanaka, and O. Teryaev at the J-PARC workshop in 2015 http://research.kek.jp/group/hadron10/j-parc-hm-2015/.
9. S. Kumano, M. Strikman, and K. Sudoh, Phys. Rev. D 80 (2009) 074003.
10. Presentations by A. Bacchetta and I. Tsutsui at the workshop, http://research.kek.jp /group/hadron10/j-parc-hm-2015/.
11. M. Anselmino *et al.*, Phys. Rev. D 70 (2004) 074025 [Figure 9 is printed with permission (©2004 APS)]; U. D'Alesio and F. Murgia, AIP Conf. Proc. 915 (2007) 559.
12. A. Airapetian *et al.* (HERMES Collaboration) Phys. Rev. Lett. 95 (2005) 242001; F. E. Close and S. Kumano, Phys. Rev. D 42 (1990) 2377; S. Hino and S. Kumano, Phys. Rev. D 59, 094026 (1999); 60, 054018 (1999); S. Kumano, Phys. Rev. D 82 (2010) 017501; J. Phys. Conf. Series 543 (2014) 012001.
13. D. G. Crabb *et al.* Phys. Rev. Lett. 65 (1990) 3241 [Figure 11 is printed with permission (©1990 APS)].
14. E. Leader, A. V. Sidorov, and D. B. Stamenov, Phys. Rev.D 91 (2015) 054017 [Figure 12 is printed with permission (©2015 APS)].

Spin Physics (SPIN2014)
International Journal of Modern Physics: Conference Series
Vol. 40 (2016) 1660010 (13 pages)
© The Author(s)
DOI: 10.1142/S2010194516600107

Searches for the Role of Spin and Polarization in Gravity: A Five-Year Update

Wei-Tou Ni

Center for Gravitation and Cosmology, Department of Physics,
National Tsing Hua University, Hsinchu, Taiwan 30013, ROC
weitou@gmail.com

Published 29 February 2016

Searches for the role of spin in gravitation dated before the firm establishment of the electron spin in 1925. Since mass and spin, or helicity in the case of zero mass, are the Casimir invariants of the Poincaré group and mass participates in universal gravitation, these searches are natural steps to pursue. In this update, we report on the progress on this topic in the last five years after our last review. We begin with how is Lorentz/Poincaré group in local physics arisen from spacetime structure as seen by photon and matter through experiments/observations. The cosmic verification of the Galileo Equivalence Principle for photons/electromagnetic wave packets (Universality of Propagation in spacetime independent of photon energy and polarization, i.e. nonbirefringence) constrains the spacetime constitutive tensor to high precision to a core metric form with an axion degree and a dilaton degree of freedom. Hughes-Drever-type experiments then constrain this core metric to agree with the matter metric. Thus comes the metric with axion and dilation. In local physics this metric gives the Lorentz/Poincaré covariance. Constraints on axion and dilaton from polarized/unpolarized laboratory/astrophysical/cosmic experiments/observations are presented. In the end, we review the theoretical progress on the issue of gyrogravitational ratio for fundamental particles and the experimental progress on the measurements of possible long range/intermediate range spin-spin, spin-monopole and spin-cosmos interactions.

Keywords: Spin; polarization; gravity; equivalence principle; electromagnetism; spacetime structure.

PACS numbers: 04.20.Cv, 04.50.Kd, 04.60.Bc, 04.80.Nn, 11.30.Cp, 13.88.+e, 14.80.Va

1. Introduction

Both electroweak and strong interactions are strongly spin-dependent. In searching for the role of spin in gravitation, we look into the empirical foundations of current theories of gravitation, i.e. general relativity and other relativistic theories of gravity.

Relativity sprang out from Maxwell-Lorentz theory of electromagnetism. Therefore we first look into the empirical role of polarization and spin in the gravity-coupling of electromagnetism; we look into photon/electromagnetic wave packet propagation in spacetime. Astrophysical observations and cosmological observations have shown that our early universe evolves according to Hot Big Bang theory. Particle physics experiments have established the Standard Model. ATLAS and CMS experiments in LHC (Large Hadron Collider) have discovered the Higgs particle and pushed the validity of the Standard Model to above the Higgs energy empirically. After the electroweak phase transition around Higgs energy about 100 ps since the Hot Big Bang, the electromagnetism separated out and the Maxwell-Lorentz theory and QED became valid. We look for empirical evidences for gravity-coupling of electromagnetism since then in Section 2. In Paper I,[1] we reviewed the experimental and theoretical efforts in searches for the role of spin and polarization in gravity up to the end of 2009. In the present paper, we give a five-year update. In Section 3, we review theoretical works on the gyrogravitational effects. In Section 4, we review experimental progress on the measurement of long range/intermediate range spin-spin, spin-monopole and spin-cosmos interactions. In Section 5, we look into the future.

2. Search for Polarization Effects in the Coupling of Gravity to Electromagnetism

In the genesis of general relativity, there are two important cornerstones: the Einstein Equivalence Principle (EEP) and the metric as the dynamic quantity of gravitation[2] (See also, Ref. [3]). With research activities on cosmology thriving, people have been looking actively for alternative theories of gravity again for more than thirty years. Recent theoretical studies include scalars, pseudoscalars, vectors, metrics, bimetrics, strings, loops, etc. as dynamic quantities of gravity. To find out how metric and other possible fields arises from experiments/observations, we notice that Maxwell-Lorentz electrodynamics can be put into premetric form dependent only on differential structure, not on metric/connection or other geometric structures.[4–9]

2.1. *Premetric formulation of electromagnetism*

Maxwell equations for macroscopic/spacetime electrodynamics in terms of independently measurable field strength F_{kl} ($\boldsymbol{E}$, $\boldsymbol{B}$) and excitation (density with weight +1) H^{ij} ($\boldsymbol{D}$, $\boldsymbol{H}$) do not need metric as primitive concept (See, e. g., Hehl and Obukhov [9]):

$$H^{ij}_{,j} = -4\pi J^i, \tag{1a}$$

$$e^{ijkl} F_{jk,l} = 0, \tag{1b}$$

with J^k the charge 4-current density and e^{ijkl} the completely anti-symmetric tensor density of weight +1 ($e^{0123} = 1$). We use units with the nominal light velocity parameter c equal to 1. To complete this set of equations, a constitutive relation is needed between the excitation and the field:

$$H^{ij} = \chi^{ijkl} F_{kl}. \tag{2}$$

Both H^{ij} and F_{kl} are antisymmetric, hence χ^{ijkl} must be antisymmetric in i and j, and in k and l. Therefore the constitutive tensor density χ^{ijkl} (with weight +1) has 36 independent components, and can be uniquely decomposed into principal part (P), skewon part (Sk) and axion part (Ax) as given in [9, 10]:

$$\chi^{ijkl} = {}^{(\text{P})}\chi^{ijkl} + {}^{(\text{Sk})}\chi^{ijkl} + {}^{(\text{Ax})}\chi^{ijkl}, \quad (\chi^{ijkl} = -\chi^{jikl} = -\chi^{ijlk}), \tag{3}$$

with

$${}^{(\text{P})}\chi^{ijkl} = (1/6)[2(\chi^{ijkl} + \chi^{klij}) - (\chi^{iklj} + \chi^{ljik}) - (\chi^{iljk} + \chi^{jkil})], \tag{4a}$$

$${}^{(\text{Ax})}\chi^{ijkl} = \chi^{[ijkl]} = \varphi e^{ijkl}, \tag{4b}$$

$${}^{(\text{Sk})}\chi^{ijkl} = (1/2)(\chi^{ijkl} - \chi^{klij}). \tag{4c}$$

The principal part has 20 degrees of freedom. The axion part has one degree of freedom. The Hehl-Obúkhov-Rubilar skewon part (4c) can be represented as

$${}^{(\text{Sk})}\chi^{ijkl} = e^{ijmk} S_m^l - e^{ijml} S_m^k, \tag{5}$$

with S_m^n a traceless tensor of 15 independent degrees of freedom.[9,10] If there is metric, the S_m^n can be raised or lowered with this metric; when S_{mn} is symmetric it is called Type I, and when it is antisymmetric it is called Type II.[11] For the skewonless case (i.e., $\chi^{ijkl} = \chi^{klij}$), the Maxwell equations can be derived from the Lagrangian density:

$$L_{\text{I}} = L_{\text{I}}^{(\text{EM})} + L_{\text{I}}^{(\text{EM}-\text{P})} + L_{\text{I}}^{(\text{P})} = -(1/(16\pi))\chi^{ijkl} F_{ij} F_{kl} - A_k J^k + L_{\text{I}}^{(\text{P})}, \tag{6}$$

with $L_{\text{I}}^{(\text{EM})} = -(1/(16\pi))\chi^{ijkl} F_{ij} F_{kl}$, A_k the electromagnetic potential guaranteed by (1b), J^k the 4 charge current density and $L_{\text{I}}^{(\text{P})}$ the particle Lagrangian density. The Lagragian density (6) has been used to study the equivalence principles and their empirical foundations in the 1970s and 1980s.[12-14] Photon sector of the Standard Model Extension (SME)[15] is contained in the χ^{ijkl}-framework with $L_{\text{I}}^{(\text{EM})}$ of (6).[16] In the Standard Model Supplement (SMS),[17] photon sector is different from but overlaps with the χ^{ijkl}-framework.

In macroscopic medium, the constitutive tensor gives the medium-coupling to electromagnetism; it depends on the (thermodynamic) state of the medium and, in turn, depends on temperature, pressure etc. In gravity, the constitutive tensor (2) gives the gravity-coupling to electromagnetism; it depends on the gravitational field(s) and, in turn, depends on the matter distribution and its state. Now the issue is how to arrive at the metric from the constitutive tensor through experiments/observations. That is, how to build the metric empirically and test the Einstein Equivalence Principle thoroughly. Are there other degrees of freedom to be explored?

Since ordinary energy compared to Planck energy is very small, in this situation we can assume that the gravitational (or spacetime) constitutive relation tensor is a linear and local function of gravitational field(s).

2.2. *Wave propagation and dispersion relation*

The sourceless Maxwell equation (1b) is equivalent to the local existence of a 4-potential A_i such that

$$F_{ij} = A_{j,i} - A_{i,j}, \tag{7}$$

with a gauge transformation freedom of adding an arbitrary gradient of a scalar function to A_i. The Maxwell equation (1a) in vacuum then becomes

$$(\chi^{ijkl} A_{k,l})_{,j} = 0. \tag{8}$$

Using the derivation rule, we have

$$\chi^{ijkl} A_{k,l,j} + \chi^{ijkl}_{,j} A_{k,l} = 0. \tag{9}$$

Neglecting $\chi^{ijkl}_{,m}$ for slowly varying/nearly homogeneous field/medium, or in the lowest eikonal approximation, (9) becomes

$$\chi^{ijkl} A_{k,lj} = 0. \tag{10}$$

In the weak field or dilute medium, we assume

$$\chi^{ijkl} = \chi^{(0)ijkl} + \chi^{(1)ijkl} + O(2), \tag{11}$$

where $O(2)$ means second order in $\chi^{(1)}$. Since the deviation/violation from the Einstein Equivalence Principle would be small, in the following we assume that

$$\chi^{(0)ijkl} = (1/2)g^{ik}g^{jl} - (1/2)g^{il}g^{kj}, \tag{12}$$

and that $\chi^{(1)ijkl}$ is small compared with $\chi^{(0)ijkl}$. We can then find a locally inertial frame such that g^{ij} becomes the Minkowski metric η^{ij} good to the derivative of the metric. To look for wave solutions, we use eikonal approximation and choose z-axis in the wave propagation direction so that the solution takes the following form:

$$A = (A_0, A_1, A_2, A_3)e^{ikz - i\omega t}. \tag{13}$$

We expand the solution as

$$A_i = [A_i^{(0)} + A_i^{(1)} + O(2)]e^{ikz - i\omega t}. \tag{14}$$

Imposing radiation gauge condition in the zeroth order in the weak field/dilute medium/weak EEP violation approximation, we find the zeroth order solution of

(14) and the zeroth order dispersion relation first and then we derive the dispersion relation to first order in Ref. [11]:

$$\omega = k[1 + 1/2(A_{(1)} + A_{(2)}) \pm 1/2((A_{(1)} - A_{(2)})^2 + 4B_{(1)}B_{(2)})^{1/2}] + O(2), \quad (15)$$

with

$$A_{(1)} \equiv \chi^{(1)1010} - (\chi^{(1)1013} + \chi^{(1)1310}) + \chi^{(1)1313}, \qquad (16a)$$

$$A_{(2)} \equiv \chi^{(1)2020} - (\chi^{(1)2023} + \chi^{(1)2320}) + \chi^{(1)2323}, \qquad (16b)$$

$$B_{(1)} \equiv \chi^{(1)1020} - (\chi^{(1)1023} + \chi^{(1)1320}) + \chi^{(1)1323}, \qquad (16c)$$

$$B_{(2)} \equiv \chi^{(1)2010} - (\chi^{(1)2013} + \chi^{(1)2310}) + \chi^{(1)2313}. \qquad (16d)$$

From (15) the group velocity is

$$v_g = \partial\omega/\partial k = 1 + 1/2(A_{(1)} + A_{(2)}) \pm 1/2((A_{(1)} - A_{(2)})^2 + 4B_{(1)}B_{(2)})^{1/2} + O(2). \qquad (17)$$

We note that $A_{(1)}$ and $A_{(2)}$ contain only the principal part of χ; $B_{(1)}$ and $B_{(2)}$ contain only the principal and skewon part of χ. The axion part drops out and does not contribute to the dispersion relation in the eikonal approximation. The principal part $^{(\mathrm{P})}B$ and skewon part $^{(\mathrm{Sk})}B$ of $B_{(1)}$ are as follows:

$$^{(\mathrm{P})}B = (1/2)(B_{(1)} + B_{(2)}); \quad ^{(\mathrm{Sk})}B = (1/2)(B_{(1)} - B_{(2)}), \qquad (18a)$$

$$B_{(1)} = {}^{(\mathrm{P})}B + {}^{(\mathrm{Sk})}B; \quad B_{(2)} = {}^{(\mathrm{P})}B - {}^{(\mathrm{Sk})}B. \qquad (18b)$$

The quantity under the square root sign is

$$\xi \equiv (A_{(1)} - A_{(2)})^2 + 4B_{(1)}B_{(2)} = (A_{(1)} - A_{(2)})^2 + 4(^{(\mathrm{P})}B)^2 - 4(^{(\mathrm{Sk})}B)^2. \qquad (19)$$

Depending on the sign or vanishing of ξ, we have the following three cases of electromagnetic wave propagation:

(i) $\xi > 0$, $(A_{(1)} - A_{(2)})^2 + 4(^{(\mathrm{P})}B)^2 > 4(^{(\mathrm{Sk})}B)^2$: There is birefringence of wave propagation;

(ii) $\xi = 0$, $(A_{(1)} - A_{(2)})^2 + 4(^{(\mathrm{P})}B)^2 = 4(^{(\mathrm{Sk})}B)^2$: There are no birefringence and no dissipation/amplification in wave propagation;

(iii) $\xi < 0$, $(A_{(1)} - A_{(2)})^2 + 4(^{(\mathrm{P})}B)^2 < 4(^{(\mathrm{Sk})}B)^2$: There is no birefringence, but there are both dissipative and amplifying modes in wave propagation.

In Ref. [11], we have shown that for $B_{(1)} = B_{(2)}$ (i.e., $^{(\mathrm{Sk})}B = 0$), the non-birefringence condition (Galileo Equivalence Principle for photons/electromagnetic wave packets) for wave propagation in all directions implies the constitutive tensor can be put into the following form:

$$\chi^{ijkl} = {}^{(\mathrm{P})}\chi^{(1)ijkl} + {}^{(\mathrm{Ax})}\chi^{(1)ijkl} + {}^{(\mathrm{SkII})}\chi^{(1)ijkl}$$

$$= 1/2(-h)^{1/2}[h^{ik}h^{jl} - h^{il}h^{kj}]\psi + \varphi e^{ijkl} + 1/2(-\eta)^{1/2}$$

$$\times (p^{ik}\eta^{jl} - p^{il}\eta^{jk} + \eta^{ik}p^{jl} - \eta^{il}p^{jk}), \qquad (20)$$

to first-order in terms of $h^{(1)ij}$, ψ, φ, and p^{ij} with the fields $h^{(1)ij}$, $(\psi - 1)$, φ, and p^{ij} defined by appropriate expressions of $\chi^{(1)}$s ($h^{ij} \equiv \eta^{ij} + h^{(1)ij}$, $h \equiv \det h_{ij}$). In the skewonless case, the nonbirefringence condition implies that the constitutive tensor is of the form

$$\chi^{ijkl} = {}^{(P)}\chi^{(1)ijkl} + {}^{(A)}\chi^{(1)ijkl} = 1/2(-h)^{1/2}[h^{ik}h^{jl} - h^{il}h^{kj}]\psi + \varphi e^{ijkl}, \quad (21)$$

as reviewed in Paper I.[1]

To derive the influence of the axion field and the dilaton field of the constitutive tensor (21) on the dispersion relation, one needs to keep the second term in equation (9). This has been done for the axion field in references [1, 18–22], and for the joint dilaton field and axion field in Ref. [23]. Near the origin in a local inertial frame, the dispersion relation in dilaton field ψ and axion field φ is

$$\omega = k - (i/2)\psi^{-1}(\psi_{,0} + \psi_{,3}) \pm \psi^{-1}(\varphi_{,0} + \varphi_{,3}) + O(2), \quad (22)$$

for plane wave propagating in the z-axis direction. The group velocity is $v_g = \partial\omega/\partial k = 1$; there is no birefringence. For plane wave propagating in direction $n^\mu = (n^1, n^2, n^3)$ with $(n^1)^2 + (n^2)^2 + (n^3)^2 = 1$, the solution is

$$\begin{aligned}
A(n^\mu) &\equiv (A_0, A_1, A_2, A_3) = (0, \underline{A}_1, \underline{A}_2, \underline{A}_3)\exp(-ikn^\mu x_\mu - i\omega t) \\
&= (0, \underline{A}_1, \underline{A}_2, \underline{A}_3)\exp[-ikn^\mu x_\mu - ikt \pm (-i)\psi^{-1}(\varphi_{,0}\,t - n^\mu\varphi_{,\mu}\,n_\nu x^\nu) \\
&\quad -(1/2)\psi^{-1}(\psi_{,0}\,t + n^\mu\psi_{,\mu}\,n_\nu x^\nu)],
\end{aligned} \quad (23)$$

where $\underline{A}_\mu = \underline{A}_\mu^{(0)} + n_\mu n^\nu \underline{A}_\nu^{(0)}$ with $\underline{A}_1^{(0)} = \pm i\underline{A}_2^{(0)}$ and $\underline{A}_3^{(0)} = 0$ [$n_\mu \equiv (-n^1, -n^2, -n^3)$]. There are polarization rotation for linearly polarized light due to axion field gradient, and amplification/attenuation due to dilaton field gradient.

2.3. *Empirical constraints on the spacetime constitutive tensor*

Nonbirefringence (no splitting, no retardation) for electromagnetic wave propagation independent of polarization and frequency (energy) in all directions can be formulated as a statement of Galilio Equivalence Principle for photons. However, the complete agreement with EEP for photon sector requires in addition: (i) no polarization rotation; (iii) no amplification/no attenuation in spacetime propagation; (iii) no spectral distortion. With nonbirefringence, any skewonless spacetime constitutive tensor must be of the form (21), hence no spectral distortion. From (23), (ii) and (iii) implies that the dilaton ψ and axion φ must be constant, i.e. no varying dilaton field and no varying axion field; the EEP for photon sector is observed; the spacetime constitutive tensor is of metric-induced form. Thus we tie the three observational conditions to EEP and to metric-induced spacetime constitutive tensor in the photon sector. The three observational constraints are reviewed in the following 3 sub-subsections with accuracies summarized in Table 1. In Section 2.3.4, we discuss the skewonful case.

Table 1. Constraints on the spacetime constitutive tensor χ^{ijkl} and construction of the spacetime structure (metric + axion field φ + dilaton field ψ) from experiments/observations in skewonless case (U: Newtonian gravitational potential). g_{ij} is the particle metric.

Experiment	Constraints	Accuracy
Pulsar Signal Propagation		10^{-16}
Radio Galaxy Observation	$\chi^{ijkl} \to \frac{1}{2}\,(-h)^{1/2}[h^{ik}\,h^{jl} - h^{il}\,h^{kj}]\psi + \varphi e^{ijkl}$	10^{-32}
Gamma Ray Burst (GRB)		10^{-38}
CMB Spectrum Measurement	$\psi \to 1$	8×10^{-4}
Cosmic Polarization Rotation Experiment	$\varphi - \varphi_0 \ (\equiv \alpha) \to 0$	$\|\langle\alpha\rangle\| < 0.02,$ $\langle(\alpha-\langle\alpha\rangle)^2\rangle^{1/2} < 0.03$
Eötvös-Dicke-Braginsky Experiments	$\psi \to 1$ $h_{00} \to g_{00}$	$10^{-10}\,U$ $10^{-6}\,U$
Vessot-Levine Redshift Experiment	$h_{00} \to g_{00}$	$1.4 \times 10^{-4}\,\Delta U$
Hughes-Drever-type Experiments	$h_{\mu\nu} \to g_{\mu\nu}$ $h_{0\mu} \to g_{0\nu}$ $h_{00} \to g_{00}$	10^{-24} 10^{-19} -10^{-20} 10^{-16}

2.3.1. *Birefringence constraint*

Empirically, the nonbirefringence condition is verified by the pulsar signal propagation, the polarization observations on radio galaxies and the gamma ray burst observations.[3,24] The accuracy of verification of the nonbirefringence condition is good up to 10^{-38}.

2.3.2. *Constraints on the cosmic polarization rotation and the cosmic axion field*

From (23), for the right circularly polarized electromagnetic wave, the propagation from a point P_1 (4-point) to another point P_2 adds a phase of $\alpha = \varphi(P_2) - \varphi(P_1)$ to the wave; for left circularly polarized light, the added phase will be opposite in sign.[18] Linearly polarized electromagnetic wave is a superposition of circularly polarized waves. Its polarization vector will then rotate by an angle α. In the global situation, it is the property of (pseudo)scalar field that when we integrate along light (wave) trajectory the total polarization rotation (relative to no φ-interaction) is again $\alpha = \Delta\varphi = \varphi(P_2) - \varphi(P_1)$ where $\varphi(P_1)$ and $\varphi(P_2)$ are the values of the scalar field at the beginning and end of the wave. The constraints[1,21,25–27] listed on the axion field are from the UV polarization observations of radio galaxies and the CMB polarization observations — 0.02 for Cosmic Polarization Rotation (CPR) mean value $|\langle\alpha\rangle|$ and 0.03 for the CPR fluctuations $\langle(\alpha - \langle\alpha\rangle)^2\rangle^{1/2}$.

2.3.3. *Constraints on the dilaton field and constraints on the unique physical metric*

The amplification/attenuation induced by dilaton is independent of the frequency (energy) and the polarization of electromagnetic waves (photons). From observations, the agreement[28] with and the precise calibration of the cosmic microwave background (CMB) to blackbody radiation constrains the fractional change of dilaton $|\Delta\psi|/\psi$ to less than about 8×10^{-4} since the time of the last scattering surface of the CMB.[23] Eötvös-type experiments constrain the fractional variation of dilaton to $\sim 10^{-10}U$ where U is the dimensionless Newtonian potential in the experimental environment.[1] Vessot-Levine redshift experiment and Hughes-Drever-type experiments give further constraints.[1]

2.3.4. *Constraints on the skewon field and the asymmetric metric*

For metric principal part plus skewon part, we have shown that the Type I skewon part is constrained to $<$ a few $\times 10^{-35}$ in the weak field/weak EEP violation limit.[11] Type II skewon part is not constrained in the first order.[11] However, in the second order it induces birefringence; the nonbirefringence observations constrain the Type II skewon part to $\sim 10^{-19}$.[24] However, an additional nonmetric induced second-order contribution to the principal part constitutive tensor compensates the Type II skewon birefringence and makes it nonbirefringent.[24] This second-order contribution is just the extra piece to the (symmetric) core-metric principal constitutive tensor induced by the antisymmetric part of the asymmetric metric tensor q^{ij} (Table 2).[24]

3. Gyrogravitational Ratio

Gyrogravitational effect is defined to be the response of an angular momentum in a gravitomagnetic field produced by a gravitating source having a nonzero angular

Table 2. [24]Various 1st-order and 2nd-order effects in wave propagation on media with the core-metric based constitutive tensors. $^{(P)}\chi^{(c)}$ is the extra contribution due to antisymmetric part of asymmetric metric to the core-metric principal part for canceling the skewon contribution to birefringence/amplification-dissipation.

Constitutive tensor	Birefringence (in the geometric optics approximation)	Dissipation/ amplification	Spectro-scopic distortion	Cosmic polarization rotation
Metric: $\frac{1}{2}(-h)^{1/2}[h^{ik}h^{jl} - h^{il}h^{kj}]$	No	No	No	No
Metric + dilaton: $\frac{1}{2}(-h)^{1/2}[h^{ik}h^{jl} - h^{il}h^{kj}]\psi$	No (to all orders in the field)	Yes (due to dilaton gradient)	No	No
Metric + axion: $\frac{1}{2}(-h)^{1/2}[h^{ik}h^{jl} - h^{il}h^{kj}] + \varphi e^{ijkl}$	No (to all orders in the field)	No	No	Yes (due to axion gradient)
Metric + dilaton + axion: $\frac{1}{2}(-h)^{1/2}[h^{ik}h^{jl} - h^{il}h^{kj}]\psi + \varphi e^{ijkl}$	No (to all orders in the field)	Yes (due to dilaton gradient)	No	Yes (due to axion gradient)
Metric + type I skewon	No to first order	Yes	Yes	No
Metric + type II skewon	No to first order; yes to 2nd order	No to first order and to 2nd order	No	No
Metric + $^{(P)}\chi^{(c)}$+ type II skewon	No to first order; no to 2nd order	No to first order and to 2nd order	No	No
Asymmetric metric induced: $\frac{1}{2}(-q)^{1/2}(q^{ik}q^{jl} - q^{il}q^{jk})$	No (to all orders in the field)	No	No	Yes (due to axion gradient)

momentum. Ciufolini and E. C. Pavlis[29] have measured and verified this effect with 10–30% accuracy for the dragging of the orbit plane (orbit angular momentum) of a satellite (LAGEOS) around a rotating planet (earth) predicted for general relativity by Lense and Thirring.[30] Gravity Probe B[31] has measured and verified the dragging of spin angular momentum of a rotating quartz ball predicted by Schiff[32] for general relativity with 19% accuracy. GP-B experiment has also verified the Second Weak Equivalence Principle (WEP II) for macroscopic rotating bodies to ultra-precision.[33]

Just as in electromagnetism, we can define gyrogravitational factor as the gravitomagnetic moment (response) divided by angular momentum for gravitational interaction. We use macroscopic (spin) angular momentum in GR as standard, its gyrogravitational ratio is 1 by definition. In Ref. [34], we use coordinate transformations among reference frames to study and to understand the Lense-Thirring effect of a Dirac particle. For a Dirac particle, the wave-function transformation operator from an inertial frame to a moving accelerated frame is obtained. According to equivalence principle, this gives the gravitational coupling to a Dirac particle. From this, the Dirac wave function is solved and its change of polarization gives the gyrogravitational ratio 1 from the first-order gravitational effects. In Teryaev's talk on Spin-gravity Interactions and Equivalence Principle, he has reported his work with Obukhov and Silenko[35] on the direct calculation of the response of the spin of a Dirac particle in gravitomagnetic field and showed that it is the same as the response of a macroscopic spin angular momentum in general relativity (See, also, Tseng [36]). Randono has showed that the active frame-dragging of a polarized Dirac particle is the same as that of a macroscopic body with equal angular momentum.[37] All these results are consistent with EEP and the principle of action-equal-to-reaction. However, these findings do not preclude that the gyrogravitational ratio to be different from 1 in various different theories of gravity, notably torsion theories and Poincaré gauge theories.

What would be the gyrogravitational ratios of actual elementary particles? If they differ from one, they will definitely reveal some inner gravitational structures of elementary particles, just as different gyromagnetic ratios reveal inner electromagnetic structures of elementary particles. These findings would then give clues to the microscopic origin of gravity.

Promising methods to measure particle gyrogravitational ratio include[1]: (i) using spin-polarized bodies (e.g. polarized solid He^3, Dy-Fe, Ho-Fe, or other compounds) instead of rotating gyros in a GP-B type experiment to measure the gyrogravitational ratio of various substances; (ii) atom interferometry; (iii) nuclear spin gyroscopy; (iv) superfluid He^3 gyrometry. Notably, there have been great developments in atom interferometry[38] and nuclear gyroscopy.[39] However, to measure particle gyrogravitational ratios the precision is still short by several orders and more developments are required.

4. Search for Long Range /Intermediate Range Spin-Spin, Spin-Monopole and Spin-Cosmos Interactions

4.1. *Spin-spin experiments*

Geomagnetic field induces electron polarization within the Earth. Hunter *et al.*[40] estimated that there are on the order of 10^{42} polarized electrons in the Earth compared to $\sim 10^{25}$ polarized electrons in a typical laboratory. For spin-spin interaction, there is an improvement in constraining the coupling strength of the intermediate vector boson in the range greater than about 1 km.[40]

4.2. *Spin-monopole Experiments*

In Paper I, we have used axion-like interaction Hamiltonian

$$H_{int} = [\hbar(g_s g_p)/8\pi mc](1/\lambda r + 1/r^2)\exp(-r/\lambda)\sigma \cdot \widehat{r}, \qquad (24)$$

to discuss the experimental constraints on the dimensionless coupling $g_s g_p/\hbar c$ between polarized (electron) and unpolarized (nucleon) particles. In (24), λ is the range of the interaction, g_s and g_p are the coupling constants of vertices at the polarized and unpolarized particles, m is the mass of the polarized particle and σ is Pauli matrix 3-vector. Hoedll *et al.*[41] have pushed the constraint to shorter range by about one order of magnitude since our last review. In this update, we see also good progress in the measurement of spin-monopole coupling between polarized neutrons and unpolarized nucleons.[42–44] Tullney *et al.* obtained the best limit on this coupling for force ranges between 3×10^{-4} m and 0.1 m.

4.3. *Spin-cosmos experiments*

For the analysis of spin-cosmos experiments for elementary particles, one usually uses the following Hamiltonian:

$$H_{cosmic} = C_1\sigma_1 + C_2\sigma_2 + C_3\sigma_3, \qquad (25)$$

in the cosmic frame of reference for spin half particle with C's constants and σ's the Pauli spin matrices (see, e.g. [45] or Paper I). The best constraint now is on bound neutron from a free-spin-precession ^{3}He–^{129}Xe comagnetometer experiment performed by Allmendinger *et al.*[39] The experiment measured the free precession of nuclear spin polarized ^{3}He and ^{129}Xe atoms in a homogeneous magnetic guiding field of about 400 nT. As the laboratory rotates with respect to distant stars, Allmendinger *et al.* looked for a sidereal modulation of the Larmor frequencies of the collocated spin samples due to (25) and obtained an upper limit of 8.4×10^{-34} GeV (68% C.L.) on the equatorial component $C^n_\perp$ for neutron. This constraint is more stringent by 3.7×10^4 fold than the limit on that for electron.[46] Using a ^{3}He-K co-magnetometer, Brown *et al.*[47] constrained $C^p_\perp$ for the proton to be less than 6×10^{-32} GeV.

5. Outlook

Polarization and spin are important in verifying Galileo Equivalence Principle and Einstein Equivalence Principle which are important cornerstones of general relativity and metric theories of gravity. General relativity and relativistic theories of gravity are bases for modern cosmology. It is not surprising that cosmological observations on polarization phenomena become the ultimate test ground of the equivalence principles, especially for the photon sector. Some of the dispersion relation tests are reaching second order in the ratio of Higgs boson mass and Planck mass. Ultra-precise laboratory experiments are reaching ground in advancing constraints on various (semi-)long-range spin interactions. Sooner or later, experimental efforts will reach the precision of measuring the gyrogravitational ratios of elementary particles. All these developments may facilitate ways to explore the origins of gravity.

Acknowledgments

We would like to thank F. Allmendinger, C. Fu, H. Gao, B.-Q. Ma, Yu. N. Obukhov, O. Teryaev, K. Tullney for helpful discussions. We would also like to thank the National Science Council (Grant No. NSC102-2112-M-007-019) and the National Center for Theoretical Sciences (NCTS) for supporting this work in part.

References

1. W.-T. Ni, *Rep. Prog. Phys.* **73**, 056901 (2010) [Paper I].
2. Jürgen Renn, (Ed.) *The genesis of general relativity: Sources and Interpretations*, in 4 volumes (Boston Studies in the Philosophy and History of Science) (Springer, 2007).
3. W.-T. Ni, Equivalence principles, spacetime structure and the cosmic connection, to be published as Chapter 5 in the book: *One Hundred Years of General Relativity: from Genesis and Empirical Foundations to Gravitational Waves, Cosmology and Quantum Gravity*, edited by W.-T. Ni (World Scientific, Singapore, 2015).
4. A. Einstein, Eine Neue Formale Deutung der Maxwellschen Feldgleichungen der Elektrodynamik, *Königlich Preußische Akademie der Wissenschaften* (Berlin), 184–188 (1916); See also, A new formal interpretation of Maxwell's field equations of Electrodynamics, in *The Collected Papers of Albert Einstein*, Vol. 6, A. J. Kox *et al.*, eds. (Princeton University Press: Princeton, 1996), pp. 263–269.
5. H. Weyl, *Raum-Zeit-Materie*, Springer, Berlin, 1918; See also, *Space-Time-Matter*, English translation of the 4th German edition of 1922 (Dover, Mineola, New York, 1952).
6. F. Murnaghan, *Phys. Rev.* **17**, 73–88 (1921).
7. F. Kottler, Maxwell'sche Gleichungen und Metrik. *Sitzungsber. Akad. Wien IIa* **131**, 119–146 (1922).
8. É. Cartan, Sur les variétés à connexion affine et la Théorie de la Relativité Généralisée, *Annales scientifirues de l'École Normale Supérieure* **40**, pp. 325–412, **41**, pp. 1–25, **42**, pp. 17–88 (1923/1925); See also, *On Manifolds with an Affine Connection and the Theory of General Relativity*, English translation of the 1955 French edition. Bibliopolis, Napoli, 1986.
9. F. W. Hehl and Yu. N. Obukhov, *Foundations of Classical Electrodynamics: Charge, Flux, and Metric* (Birkhäuser: Boston, MA, 2003).

10. F. W. Hehl, Yu. N. Obukhov and G. F. Rubilar, On a possible new type of a T-odd skewon field linked to electromagnetism, in: A. Macias, F. Uribe, E. Diaz (Eds.), *Developments in Mathematical and Experimental Physics, Volume A: Cosmology and Gravitation* (Kluwer Academic/Plenum, New York, 2002), pp. 241–256 [gr-rc/0203096].

11. W.-T. Ni, *Phys. Lett. A* **378**, 1217–1223 (2014).

12. W.-T. Ni, *Phys. Rev. Lett.* **38**, 301–304 (1977).

13. W.-T. Ni, Equivalence Principles and Precision Experiments, in *Precision Measurement and Fundamental Constants II*, ed. by B. N. Taylor and W. D. Phillips, Natl. Bur. Stand. (US) Spec. Publ. 617 (1984), pp. 647–651.

14. W.-T. Ni, Equivalence Principles, Their Empirical Foundations, and the Role of Precision Experiments to Test Them, in *Proceedings of the 1983 International School and Symposium on Precision Measurement and Gravity Experiment*, Taipei, Republic of China, January 24-February 2, 1983, ed. by W.-T. Ni (Published by National Tsing Hua University, Hsinchu, Taiwan, Republic of China, 1983), pp. 491–517 [http://astrod.wikispaces.com/].

15. D. Colladay and V. A. Kostelecký, *Phys. Rev. D*, **58**, 116002 (1998).

16. The photon sector of the SME Lagrangian is given by $L_{\text{photon}}^{\text{total}} = -(1/4)F_{\mu\nu}F^{\mu\nu} - (1/4)\,(k_F)_{\kappa\lambda\mu\nu}F^{\kappa\lambda}F^{\mu\nu} + (1/2)\,(k_{AF})^{\kappa}\varepsilon_{\kappa\lambda\mu\nu}A^{\lambda}F^{\mu\nu}$ (equation (31) of [15]). The CPT-even part $(-(1/4)\,(k_F)_{\kappa\lambda\mu\nu}F^{\kappa\lambda}F^{\mu\nu})$ has constant components $(k_F)_{\kappa\lambda\mu\nu}$ which correspond one-to-one to our χ's when specialized to constant values minus the special relativistic χ with the constant axion piece dropped, i.e. $(k_F)^{\kappa\lambda\mu\nu} = \chi^{\kappa\lambda\mu\nu} - (1/2)(\eta^{\kappa\mu}\eta^{\lambda\nu} - \eta^{\kappa\nu}\eta^{\lambda\mu})$. The CPT-odd part $(k_{AF})^{\kappa}$ also has constant components which correspond to the derivatives of axion $\varphi,^{\kappa}$ when specialized to constant values.

17. L. Zhou and B.-Q. Ma, *Mod. Phys. Lett. A* **25**, 2489 (2010) [arXiv:1009.1331].

18. W.-T. Ni, A Nonmetric Theory of Gravity, preprint, Montana State University (1973) [http://astrod.wikispaces.com/].

19. W.-T. Ni, Timing Observations of the Pulsar Propagations in the Galactic Gravitational Field as Precision Tests of the Einstein Equivalence Principle, in *Proceedings of the Second Asian-Pacific Regional Meeting of the International Astronomical Union on Astronomy, Bandung, Indonesia – 24 to 29 August 1981*, ed. by B. Hidayat and M. W. Feast (Published by Tira Pustaka, Jakarta, Indonesia, 1984), pp. 441–448.

20. W.-T. Ni, *Chin. Phys. Lett.* **22**, 33–35 (2005).

21. W.-T. Ni, *Prog. Theor. Phys. Suppl.* **172**, 49 (2008) [arXiv:0712.4082].

22. Y. Itin, *Gen. Rel. Grav.* **40**, 1219 (2008).

23. W.-T. Ni, *Phys. Lett. A* **378**, 3413–3418 (2014).

24. W.-T. Ni, Spacetime structure and asymmetric metric from the premetric formulation of electromagnetism, arXiv:1411.0460.

25. S. di Serego Alighieri, W.-T. Ni and W.-P. Pan, *Astrophys. J.* **792** (2014) 35.

26. H. H. Mei, W.-T. Ni, W.-P. Pan, L. Xu and S. di Serego Alighieri, New constraints on cosmic polarization rotation from the ACTPol cosmic microwave background B-Mode polarization observation and the BICEP2 constraint update, arXiv:1412.8569.

27. S. di Serego Alighieri, Cosmic Polarization Rotation: an Astrophysical Test of Fundamental Physics, *Int. J. Mod. Phys. D*, in press, arXiv:1501.06460.

28. D. J. Fixsen, *Astrophys. J.* **707**, 916 (2009).

29. I. Ciufolini and E. C. Pavlis, *Nature* **431**, 958 (2004).

30. J. Lense and H. Thirring, *Phys. Z.* **19**, 156 (1918).

31. C. W. F. Everitt *et al.*, *Phys. Rev. Lett.* **106**, 221101 (2011) [arXiv:1105.3456].

32. L. I. Schiff, *Phys. Rev. Lett.* **4**, 215 (1960).

33. W.-T. Ni, *Phys. Rev. Lett.* **106**, 221101 (2011) [arXiv:1105.3456].

34. Y.-C. Huang and W.-T. Ni, Propagation of Dirac Wave Functions in Accelerated Frames of Reference, arXiv:gr-qc/0407115.

35. Y. N. Obukhov, A. J. Silenko and O. V. Teryaev, *Phys. Rev. D* **88**, 084014 (2013).

36. H.-H. Tseng, *On the Equation of Motion of a Dirac Particle in Gravitational Field and its Gyro-Gravitational Ratio*, M. S. (In Chinese with an English abstract, Advisor: W.-T. Ni), National Tsing Hua University, Hsinchu, 2001, for a derivation in the weak field limit.

37. A. Randono, *Phys. Rev. D* **81**, 024027 (2010).

38. T. Schuldt *et al.*, Design of a dual species atom interferometer for space, *Exp. Astron.*, in press, arXiv:1412.2713.

39. F. Allmendinger *et al.*, *Phys. Rev. Lett.* **112**, 110801(2014).

40. L. Hunter *et al.*, *Science* **339**, 928 (2013).

41. S. A. Hoedl *et al.*, *Phys. Rev. Lett.* **106**, 100801 (2011).

42. P.-H. Chu *et al.*, *Phys. Rev. D* **87**, 011105(R) (2013).

43. M. Bulatowicz, *et al.*, *Phys. Rev. Lett.* **111**, 102001 (2013).

44. K. Tullney *et al.*, *Phys. Rev. Lett.* **111**, 100801 (2013).

45. P. R. Phillips, New tests for the invariance of the vacuum state under the Lorentz group, *Phys. Rev.* **139**, B491–B494 (1965).

46. B. R. Heckel, E. G. Adelberger, C. E. Cramer, T. S. Cook, S. Schlamminger and U. Schmidt, *Phys. Rev. D* **78**, 092006 (2008).

47. J. M. Brown, S. J. Smullin, T. W. Kornack and M. V. Romalis, *Phys. Rev. Lett.* **105**, 151604 (2010).

Spin Physics (SPIN2014)
International Journal of Modern Physics: Conference Series
Vol. 40 (2016) 1660011 (6 pages)
© The Author(s)
DOI: 10.1142/S2010194516600119

Symposium Summary

Richard G. Milner

Laboratory for Nuclear Science,
Massachusetts Institute of Technology,
Cambridge, MA 02139, USA
milner@mit.edu

Published 29 February 2016

The Stern-Gerlach experiment and the origin of electron spin are described in historical context. SPIN 2014 occurs on the fortieth anniversary of the first International High Energy Spin Physics Symposium at Argonne in 1974. A brief history of the international spin conference series is presented.

Keywords: Spin; quantum mechanics; history.

1. Introduction

In these brief, concluding remarks to this excellent scientific meeting, I have decided to recount some of the early origins of spin physics. I have been motivated to do this by the many young students at this meeting who may be less familiar with the events almost one hundred years ago on a far-off continent. Further, I believe that there are important lessons to be learned for all from an understanding of how the research frontiers are confronted and discoveries are made. It differs significantly from the way we learn and teach settled science years later from textbooks in classrooms.

In the early 1920's, the physicist's description of the atom was based on the planetary model of Bohr[1] and the quantization of the z component of angular momentum by Sommerfeld.[2] The electrons orbited the nucleus in stationary, circular paths at fixed distances from the nucleus. Electrons could gain or lose energy by jumping from one allowed orbit to another. The atom had angular momentum $L = n\hbar$, where $n = 1, 2, 3....$ is the principal quantum number. The electron in the $n = 1$ ground state had a magnetic moment of $\frac{e\hbar}{2m_e c}$, the Bohr magneton. This model successfully explained the Rydberg formula deduced from experimental observation of the spectral lines of the hydrogen atom. While the Bohr-Sommerfeld theory was the accepted description it was widely recognized that it could not be the final word.

Stern and von Laue are quoted to have declared in 1913: *"If this nonsense of Bohr should in the end prove to be right, we will quit physics!"*.

2. The Stern-Gerlach Experiment

The Stern-Gerlach experiment was carried out in 1922 in Frankfurt, Germany by Otto Stern and Walther Gerlach. Stern received a Ph.D. in physical chemistry from the University of Breslau in 1912. He was the first pupil of Albert Einstein. Importantly for our story, Stern was a cigar smoker. Walther Gerlach received his Ph.D. in experimental physics also in 1912 at the University of Tübingen. Stern had a position at the Johann Wolfgang Goethe University of Frankfurt am Main from 1915 until 1921, when he became a professor at the University of Rostock. Gerlach held a position at Frankfurt from 1920 until 1925, when he answered a call to become a professor at the University of Tübingen. Max Born was also at the University of Frankfurt from 1919 to 1921, when he left to take a professor position at the University of Göttingen.

Otto Stern conceived of the Stern-Gerlach experiment[3] one cold morning as he lay in his warm bed.[5] He was focused on experimentally demonstrating space quantization. The idea was to generate an atomic beam of silver atoms from a hot oven. In the Bohr-Sommerfeld model, the atoms are assumed to have a magnetic moment which for the ground state $n = 1$ would be $\pm \frac{e\hbar}{2m_e c}$. The atoms in the oven at temperature T would have a distribution of velocities given by the Maxwell-Boltzmann distribution so that some would escape a hole in the wall of the oven. This atomic beam would pass through an inhomogeneous magnetic field. The inhomogeneous field will exert a force on the magnetic dipole which should cause the silver atomic beam to be split. The beams are detected some distance beyond the magnet using a photographic plate. Stern assumed that $T = 1000°$ K, $\frac{\partial B}{\partial z} = 10^4$ gauss cm^{-1}, which would produce a separation on the plate of 1.12×10^{-3} cm.[3]

Stern discussed his idea with Born, who was initially unimpressed: *"I thought always that this space quantization was a kind of symbolic expression for something which you don't understand......I tried to persuade Stern that there was no sense in it but then he told me it was worth a try"*. The experiment took more than a year to realize. Securing the necessary funding was a major challenge. Having been convinced of the importance of the experiment, Born gave public lectures on Einstein and relativity and charged an entrance fee. Crucially, a check from Harry Goldman (founder of Goldman-Sachs) in New York saved the experiment.

When the experiment was first carried out and Gerlach removed the plate from the vacuum, no sign of the silver was visible.[5] Cigar-smoking Stern received the plate from Gerlach and slowly the silver became visible. Stern's cheap cigars contained sulfur and the smoke interacting with the silver produced silver sulfide, which is black and easily visible. To convince skeptics, this effect of the cigar smoke on the silver was reenacted in 2002 and successfully confirmed.[5]

The published results[4] were obtained with a magnetic field gradient up to twenty times that assumed in Stern's proposal. This greatly increased the separation to about 0.2 mm, which was visible. It is doubtful that the effect could be seen with the original value of field gradient. At the time, the success of the experiment was heralded as a crucial validation of the Bohr-Sommerfeld theory over the classical theory of the atom. It showed clearly that spatial quantization exists, a phenomenon that can be accommodated only within a quantum mechanical theory.

3. The Spinning Electron

In 1921, A.H. Compton suggested that the electron has a magnetic moment. In part, this was motivated to explain the observed, mysterious doubling of atomic states, beyond what was predicted by the Bohr-Sommerfeld quantization rules. This doubling was known as *Mechanische Zweideutigkeit* in German and as *duplexity* in English. In 1925, the Pauli Exclusion Principle was formulated[6] as: *no two electrons can have identical quantum numbers.*

Also in that year, Leiden graduate students Uhlenbeck and Goudsmit first hypothesized[7] intrinsic spin as a property of the electron. This occurred over the strong objections of some prominent physicists but with the support of their advisor, Paul Ehrenfest. In their Nature letter they write: *"It seems that the introduction of the concept of the spinning electron makes it possible throughout to maintain the principle of the successive building up of atoms utilized by Bohr in his general discussion of the relations between spectra, and the natural system of the elements. Above all, it may be possible to account for the important results arrived at by Pauli without having to assume an unmechanical duality in the binding of the electrons."* In the succeeding letter in the same journal, Bohr fully agreed.

The objections to the idea of spin by physicists of the stature of Pauli and Lorentz were not trivial. The electron was viewed as having a classical radius $r_e = \frac{1}{4\pi\epsilon_0} \cdot \frac{e^2}{m_e c^2} = 2.8 \times 10^{-15}$ m. If the electron was spinning with orbital angular momentum of 1 Bohr magneton, then the velocity at the surface of the electron significantly exceeded the speed of light. A violation of Einstein's theory of relativity was unacceptable.

In 1926, Thomas[8] correctly applied relativistic calculations to spin-orbit coupling in atomic systems and resolved a missing factor of two in the derived g-values. Also in 1926, Fermi[9] and Dirac[10] developed the Fermi-Dirac statistics for electrons. It was immediately applied to describe stellar collapse to a white dwarf,[11] to electrons in metals,[12] and to field electron emission from metals.[13]

In 1928, Dirac developed[14] his elegant equation for spin-$\frac{1}{2}$ particles. In this formulation, solutions are four-component spinors which are interpreted as positive and negative energy states of spin $\pm\frac{1}{2}$ each. Dirac predicted the existence of the positron, and the theory became the basis for the most precisely tested theory in physics, Quantum Electrodynamics. By the end of the 1920s, physicists had developed a fundamental understanding of the essential role of electron spin in explaining the electronic structure of the atom. There exist excellent, personal, historical accounts by Dirac,[15] Uhlenbeck,[16] and Goudsmit[17] of this period.

In 1927, Wrede, a student of Stern at Hamburg,[18] and Phipps and Taylor at Illinois[19] independently observed the deflection of atomic hydrogen in a magnetic field gradient. In 1929, Mott wondered[20] if electron spin can be observed directly via the scattering of electrons from atomic nuclei. Note that in the Appendix to his paper, Mott showed that the Stern-Gerlach experiment cannot be carried out for electrons. Only in 1942 did Shull *et al.* verify[21] Mott's prediction in a double scattering experiment which used 400 keV electrons from a Van de Graaf generator. In the mid-1920s, Heisenberg and Hund postulated the existence of two kinds of molecular hydrogen: *orthohydrogen* where the two proton spins are aligned parallel and *parahydrogen* where the two proton spins are antiparallel. By the end of the decade, they had been studied experimentally. Later, by deflection of orthohydrogen in a magnetic field gradient, Stern and collaborators measured the g-factor of the proton to be about 2.5 nuclear magnetons,[23] a marked deviation from the Dirac value for a pointlike spin-$\frac{1}{2}$ particle, and the first hint of its internal structure.

In the 1930s, Rabi and collaborators (inc. N. Ramsey and J. Zacharias) using molecular beams in a weak magnetic field measured the magnetic moments and nuclear spins of hydrogen, deuterium, and heavier nuclei.[24]

By the end of the 1940s, the nuclear shell model had been established.[25] This explained the properties and structure of atomic nuclei and underscored the essential role of proton and neutron spin. A key aspect was the strong role of spin-orbit coupling, which was suggested to Goeppert-Mayer by a question from Fermi.

4. The International Spin Physics Community

By the middle of the twentieth century, the intrinsic spin of subatomic particles was a cornerstone of the physicist's theoretical understanding of the fundamental structure of matter. However, spin as an experimental tool became a reality only in the 1950's, when a number of seminal experiments were carried out using spin. In 1956, Lee and Yang pointed out that parity should be violated in the weak interaction.[26] Shortly afterwards, in 1956, Wu and collaborators observed[27] parity violation in aligned ^{56}Co. In 1958, it was shown experimentally[28] using polarization techniques that the neutrino has negative helicity. 1959, the Thomas-Bargmann-Michel-Telegdi equation describing the spin precession of an electron in an external electromagnetic field was derived.[29]

In the 1960s, the discovery of pointlike constituents in the proton at SLAC using deep inelastic scattering (DIS) profoundly affected our understanding of the fundamental structure of matter. A key determination that these constituents had spin-$\frac{1}{2}$ led to their identification as the quarks of SU(3) symmetry. Important sum rules related to spin-dependent DIS were derived by Bjorken[30] and by Ellis and Jaffe.[31]

During this period, the international spin community grew significantly in size to become the active, subfield of international physics we have to-day. Beginning in 1960 at Basel, symposia on polarization phenomena in nuclear reactions were held every 5 years until 1994. Beginning in 1974 at Argonne, symposia on high energy spin physics were held every 2 years until 1994. Beginning in 1996 in Amsterdam, the international spin community became unified and a symposium on spin physics has been held every two years since then. The International Spin Physics Committee was formed to oversee the organization of this biennial symposium which has taken place here in Beijing, China in the past week. The published proceedings of these meetings form the essential record of the research activities over this time. In,[32] there is a complete tabulation of these meetings as well as references to their proceedings. Further, important conventions at Basel in 1960 for spin-$\frac{1}{2}$ particles[33] and at Madison in 1970 for spin-1 particles[34] were established to facilitate consistent discussion of spin observables.

5. Conclusion

The Stern-Gerlach experiment was the right experiment to demonstrate space quantization but was explained by the wrong theory at the time. We now know that the silver atom has an unpaired electron in the $5s$ shell and that all the other electrons are paired. The $5s$ electron is in a zero orbital angular momentum state. Thus, the inhomogeneous magnetic field exerted a force only on the magnetic dipole moment of the unpaired electron.

Two graduate students postulated spin over the strong criticism of senior physicists at the time. Their advisor fully supported them. At the time, the Stern-Gerlach experiment was not connected to spin. There is no mention of it in Uhlenbeck and Goudsmit's paper.

Neither the Stern-Gerlach experiment nor the origination of electron spin was recognized by the Nobel Prize Committee. In 1943, Stern was awarded the Nobel Prize in physics for the discovery of the magnetic moment of the proton.

Space quantization and spin are the cornerstone of the physicist's description of the universe. Consequences include: nuclear magnetic resonance, the shell model of the nucleus, optical pumping, the laser, the Lamb shift, the anomalous magnetic moments of the leptons, digital communication, atomic clocks, and the global positioning system.

I want to end by extending my warm congratulations to Profs. Haiyan Gao and Bo-Qiang Ma and their colleagues. SPIN 2014 in Beijing has been an outstanding success due to their considerable efforts. We look forward to SPIN 2016 at the University of Illinois Urbana-Champaign, USA.

Acknowledgments

I would like to acknowledge the support of my colleagues on the International Spin Physics Committee. In particular, I thank Erhard Steffens for his leadership, efficiency and untiring dedication to the stewardship of international spin physics. My research is supported by the US Department of Energy Office of Nuclear Physics under Contract Number DE-FG02-94ER40818.

References

1. N. Bohr, *Phil. Mag.*, **26**, No. 151, 1, July 1913.
2. A. Sommerfeld, *Annalen der Physik*, **51**, No. 17, 1 (1916).
3. Otto Stern, *Zeitschrift für Physik*, **7**, 249 (1921).
4. Walther Gerlach and Otto Stern, *Zeitschrift für Physik*, **9**, 353 (1922).
5. Bretislav Friedrich and Dudley Herschbach, *Physics Today*, p. 53, December 2003.
6. W. Pauli, Zeitschrift für Physik, **31**, 765 (1925).
7. G.E. Uhlenbeck and S. Goudsmit, Naturwissenchaften **13**, 953 (1925); Nature **117**, 264 (1926).
8. L.H. Thomas, Nature **117**, 514 (1926).
9. E. Fermi, Rendiconti Lincei **3**, 145 (1926).
10. P.A.M. Dirac, Proc. Roy. Soc. **A112**, 661 (1926).
11. R.H. Fowler, Monthly Notices of the Royal Astronomical Society **87**, 114 (1926).
12. A. Sommerfeld, Naturwissenschaften **15**, 824 (1927).
13. R.H. Fowler and L.W. Nordheim, Proc. Roy. Soc. **A119**, 173 (1928).
14. P.A.M. Dirac, Proc. Roy. Soc. **A117**, 610 (1928).
15. P.A.M. Dirac, *An Historical Perspective of Spin*, Proceedings of the Summer Studies on high-energy physics with polarized beams, Argonne National Laboratory, July 1974.
16. George E. Uhlenbeck, *FIFTY YEARS OF SPIN: Personal reminiscences*, Physics Today (American Institute of Physics) **29**, page 43, June 1976.
17. Samuel A. Goudsmit, *It might as well be spin*, Physics Today (American Institute of Physics) **29**, page 6, June 1976.
18. K. Wrede, Zeitschrift für Physik **41**, 569 (1927).
19. T.E. Phipps and J.B. Taylor, Phys. Rev. **29**, 309 (1927).
20. N. Mott, Proc. Roy. Soc. **A124**, 425 (1929).
21. C.G. Shull, C.T. Chase, and F.E. Myers, Phys. Rev. **63**, 29 (1943).
22. I. Estermann, R. Frisch, and O. Stern, Nature **132**, 169 (1933).
23. I. Estermann, R. Frisch, and O. Stern, Nature **132**, 169 (1933).
24. J.M.B. Kellogg, I.I. Rabi, N.F. Ramsey, Jr., J.R. Zacharias, Phys. Rev. **56**, 728 (1939).
25. M.G. Mayer, Phys. Rev. **74**, 235 (1948).
26. T.D. Lee and C.N. Yang, Phys. Rev. **104**, 822, (1956).
27. C.S. Wu, E. Ambler, R.W. Hayward, D.D. Hoppes, R.P. Hudson, Phys. Rev. **105**, 1413 (1957).
28. M. Goldhaber, L. Grodzins, and A.W. Sunyar, Phys. Rev. **109**, 1015 (1958).
29. V. Bargmann, L. Michel, and V.L. Telegdi, Phys. Rev. Lett. **2**, 435 (1959).
30. J.D. Bjorken, Phys. Rev. **148**, 1467 (1966).
31. J. Ellis and R. Jaffe, Phys. Rev. **D9**, 1444 (1974).
32. Hans Paetz gen. Schieck, *Nuclear Physics with Polarized Particles*, (Lecture Notes in Physics Vol. 842), Springer, 1st edition November 2011.
33. Proceedings of the International Symposium on Polarization Phenomena of Nucleons, P. Huber and K.P. Meyer, eds. Helvetica Phys. Acta, Suppl. VI (1961).
34. Polarization Phenomena in Nuclear Reactions, page XXV, H.H. Barschall and W. Haeberli, eds. The University of Wisconsin Press, Madison (1971).

Memorial Session

Michel Borghini and the rise of polarized targets

Colleagues and friends recall the work of a pioneer of a key technique in high-energy particle physics.

Michel's 1968 paper on his "spin-temperature model".

Michel Borghini, who passed away unexpectedly on 15 December 2012, was at CERN for more than 30 years. Born in 1934, Michel was a citizen of Monaco. He graduated from Ecole Polytechnique in 1955 and went on to obtain a degree in electrical engineering from Ecole Supérieure d'Electricité, Paris, in 1957. He then joined the group of Anatole Abragam at what was the Centre d'Etudes Nucléaires, Saclay, where he took part in the study of dynamic nuclear polarization that led to the development of the first polarized proton targets for use in high-energy physics experiments. It was here that he gained the experience that he was to develop at CERN, to the great benefit of experimental particle physics.

The basic aim with a polarized target is to line up the spins of the protons, say, in a given direction. In principle, this can be done by aligning the spins with a magnetic field but the magnitude of the proton's magnetic moment is such that it takes little energy to knock them out of alignment; thermal vibrations are sufficient. Even at low temperatures and reasonably high magnetic fields, the polarization achieved by this "brute force" method is small: only 0.1% at a temperature of 1 K and in an applied magnetic field of 1 T. To overcome this limitation, dynamic polarization exploits the much larger magnetic moment of electrons by harnessing the coupling of free proton spins in a material with nearby free electron spins. At temperatures of about 1 K, the electron spins are almost fully polarized in an external magnetic field of 1 T and the application of microwaves of around 70 GHz induces resonant transitions between the spin levels of coupled electron–proton pairs. The effect is to increase the natural, small proton polarization by more than two orders of magnitude. The polarization can be reversed with a slight change of the microwave frequency, with no need to reverse the external magnetic field.

First experiments

In 1962, Abragam's group, including Michel, reported on what was the first experiment to measure the scattering of polarized protons – in this case a 20 MeV beam derived from the cyclotron at Saclay – off a polarized proton target (Abragam *et al.* 1962). The target was a single crystal of lanthanum magnesium nitrate

$(La_2Mg_3(NO_3)_{12}.24H_2O$ or LMN), with 0.2% of the La^{3+} replaced with Ce^{3+}, yielding a proton polarization of 20%.

Michel moved to CERN three years later, where he and others from Saclay and CERN had just tested a polarized target in an experiment on proton–proton scattering at 600 MeV at the Synchrocyclotron (SC) (*CERN Courier* December 2007 p12). Developed by the Saclay group for the higher energy beams of the Proton Synchrotron (PS), the target consisted of a crystal of LMN 4.5 cm long with transverse dimensions 1.2 cm × 1.2 cm and doped with 1% neodymium. It was cooled to around 1 K in a ⁴He cryostat built in Saclay by Pierre Roubeau, in the field of a 1.8 T magnet designed by CERN's Guido Petrucci and built in the SC workshop. This target, with an average polarization of around 70%, was used in several experiments at the PS between 1965 and 1968, in both pion and proton beams with momenta of several GeV/*c*. These experiments measured the polarization parameter for $\pi^\pm$ elastic scattering and for the charge-exchange reaction $\pi^- p \rightarrow \pi^0 n$ at small values of *t*, the square of the four-momentum transfer, typically, |t| < 1 GeV².

In LMN crystals, the fraction of free, polarized protons is only around 1/16 of the total number of target protons. As a consequence, the unpolarized protons bound in the La, Mg, N and O nuclei formed a serious background in these early experiments. This background was reduced by imposing on the final-state particles the strict kinematic constraints expected from the collisions off protons at rest; the residual background was then subtracted by taking special data with a "dummy" target containing no free protons.

Michel's group at CERN thus began investigating the possibility of developing polarized targets with a higher content of free protons. In this context, in 1968 Michel published two important ▷

27

Tribute

papers in which he proposed a new phenomenological model of dynamic nuclear polarization: the "spin-temperature model" (Borghini 1968a and 1968b). The model suggested that sizable proton polarizations could be reached in frozen organic liquids doped with paramagnetic radicals. Despite some initial scepticism, in 1969 Michel's team succeeded in measuring a polarization of around 40% in a 5 cm^3 sample consisting of tiny beads made from a frozen mixture of 95% butanol (C_4H_9OH) and 5% water saturated with the free-radical porphyrexide. The beads were cooled to 1 K in an external magnetic field of 2.5 T and the fraction of free, polarized protons in the sample was around 1/4 – some four times higher than in LMN (Mango, Runólfsson and Borghini 1969).

The group at CERN went on to study a large number of organic materials doped with free-paramagnetic radicals, searching for the optimum combination for polarized targets. In this activity, where cryostats based on ^{3}He–^{4}He dilution capable of reaching temperatures below 0.1 K were developed, Michel guided two PhD students: Wim de Boer of the University of Delft (now professor at the Karlsruhe Institute of Technology) and Tapio Niinikoski of the University of Helsinki, who went on to join CERN in 1974. They finally obtained polarizations of almost 100% in samples of propanediol ($C_3H_8O_2$) doped with chromium (V) complexes and cooled to 0.1 K, in a field of 2.5 T, with 19% free, polarized protons.

In this work, the concept of spin temperature that Michel had proposed was verified by polarizing several nuclei simultaneously in a special sample containing ^{13}C and deuterons. The nuclei had different polarizations but their values corresponded to a single spin temperature in the Boltzmann formula giving the populations of the various spin states.

These targets were used in a number of experiments at CERN, at both the PS and the Super Proton Synchrotron (SPS). They measured polarization parameters in the elastic scattering of pions, kaons and protons on protons in the forward diffraction region and at backward scattering angles; in the charge-exchange reactions $K^-p \rightarrow \overline{K}^0n$ and $\overline{p}p \rightarrow \overline{n}n$; in the reaction $\pi^-p \rightarrow K^0\Lambda^0$; and in proton–deuteron scattering. In all of these experiments, Jean-Michel Rieubland of CERN provided invaluable help to ensure a smooth operation of the targets.

In the early 1970s, Michel also initiated the development of "frozen spin" targets. In these targets, the proton spins were first dynamically polarized in a high, uniform magnetic field, and then cooled to a low enough temperature so that the spin-relaxation rate of the protons would be slow even in lower magnetic fields. The targets could then be moved to the detector, thus providing more freedom in the choice of magnetic spectrometers and orientations of the polarization vector. The first frozen spin target was successfully operated at CERN in 1974.

In 1969, Michel took leave from CERN to join the Berkeley group led by Owen Chamberlain working at SLAC, where he took part in a test of T-invariance in inelastic e$^\pm$ scattering from polarized protons in collaboration with the SLAC group led by Richard Taylor. The target, built at Berkeley, was made of butanol and the SLAC 20 GeV spectrometer served as the electron (and positron) analyser. The experiment measured the up–down asymmetry for transverse target spin for both electrons and positrons. No time-reversal violations were seen at the few per cent level.

Michel with a frozen-spin target at CERN in 1976.

The target's internal structure.

Michel during his time with the CERN Staff Association.

Michel took leave to work at SLAC again in 1977, this time on a search for parity violation in deep-inelastic scattering of polarized electrons off an unpolarized deuterium target. Here, he worked on the polarized electron source and its associated laser, as well as on the electron spectrometer. The small parity-violation effects expected from the interference of the photon and Z exchanges were, indeed, observed and published in 1978. Michel then moved to the University of Michigan at Ann Arbor, where he joined the group led by Alan Krisch and took part in an experiment to measure proton–proton elastic scattering using both a polarized target and a 6 GeV polarized beam from the 12 GeV Zero Gradient Synchrotron at Argonne National Laboratory.

Michel left CERN's polarized target group in 1978, succeeded by Niinikoski. Writing in 1985 on major contributions to spin physics, Chamberlain listed the people that he felt to be "the heroes – the people who have given [this] work a special push" (Chamberlain 1985). Michel is the only one that he cites twice: with Abragam and colleagues for the first polarized target and the first experiment to use such a target; and with Niinikoski, for their introduction of the

28

Tribute

frozen spin target and showing the advantages of powerful (dilution) refrigerators. Today, polarized targets with volumes of several litres and large ^{3}He–^{4}He dilution cryostats are still in operation, for example in the NA58 (COMPASS) experiment at the SPS, where the spin structure of the proton has been studied using deep-inelastic scattering of high-energy muons (*CERN Courier* July/August 2006 p15 and September 2010 p34). Dynamic nuclear polarization has also found applications in medical magnetic-resonance imaging and Michel's spin-temperature model is still widely used.

In the 1980s, Michel took part in the UA2 experiment at CERN's SPS proton–antiproton collider, where he contributed to the calibration of the 1444 analogue-to-digital converters (ADCs) that were used to measure the energy deposited in the central calorimeter. He wrote all of the software to drive the large number of precision pulse-generators that monitored the ADC stability during data-taking.

From 1983 to 1996, he was a member of the Executive Committee of the CERN Staff Association, being its vice-president until 1990 and then its president until June 1996. After retiring from CERN in January 1999, he returned to Monaco where in 2003 he was nominated Permanent Representative of Monaco to the United Nations (New York), a post that he kept until 2005.

Michel was an outstanding physicist, equally at ease with theory and being in the laboratory. He had broad professional competences, a sharp, analytical mind, imagination and organizational skills. He is well remembered by his collaborators for his wisdom and advice, and also for his quiet demeanour and his keen but often subtle, sense of humour. His culture and interests extended well beyond science. He was also a talented tennis player. He will be sorely missed by those who had the privilege of working with him, or of being among his friends. Much sympathy goes to his two daughters, Anne and Isabelle, and to their families.

● **Further reading**

A Abragam *et al.* 1962 *Phys. Lett.* **2** 310.
M Borghini 1968a *Phys. Lett.* **26A** 242.
M Borghini 1968b *Phys. Rev. Lett.* **20** 419.
O Chamberlain 1985 *J. Phys. Colloques 46* **C2** 743.
S Mango, Ö Runólfsson and M Borghini 1969 *Nucl. Instr. Meth.* **72** 45.

Résumé
French headline

Résumé
Résumé
Résumé
Résumé
Résumé
Résumé
Résumé
Résumé
Résumé
Résumé
Résumé

His colleagues and friends, at CERN and elsewhere.

29

Spin Physics (SPIN2014)
International Journal of Modern Physics: Conference Series
Vol. 40 (2016) 1660114 (9 pages)
© The Author(s)
DOI: 10.1142/S2010194516601149

Dawn of High Energy Spin Physics — In Memory of Michel Borghini

Akira Masaike

Professor Emeritus of Kyoto University
4-3-14 Aoyamadai, Abiko-shi,
Chiba-ken, 270-1175, Japan
masaike@mvb.biglobe.ne.jp

Published 29 February 2016

High energy spin physics with the polarized proton target in 1960s is shown. The dynamic polarization in which the electronic polarization is transferred to protons in paramagnetic material by means of magnetic coupling was proposed at the beginning of 1960s. The first N-N experiment using a polarized proton target was performed with the crystal of $La_2Mg_3 (NO_3)_{12} 24H_2O$ at CEN-Saclay and Berkeley in 1962, followed by π-p experiments in several laboratories. Protons in organic materials were found to be polarized up to 80% in ^{3}He cryostats in 1969. It was helpful for large background experiments. High proton polarization was interpreted in the spin temperature theory. Spin frozen targets were constructed in early 1970s and used for experiments which require wide access angle. Michel Borghini was a main player for almost all the above works.

1. Introduction

It is a great honor for me to give a talk on the dawn of high energy spin physics in memory of Michel Borghini, who played an important role in progress of high energy spin physics. His contribution in this field has been highly appreciated.

M. Borghini was born in 1934. After graduation from Ecole Polytechnique in Paris in 1955, he joined the group of Anatole Abragam at CEN-Saclay. He started his academic career in studying the nuclear magnetic resonance in Saclay, in particular dynamic polarization of protons, which was applied to the wide field of particle physics. He devoted his full effort to the development of the polarized proton target, with which he made a significant contribution to high energy physics throughout 1960s.

M. Borghini opened a new field in particle physics, 'high energy spin physics', dawn of which I will briefly look at.

2. Dynamic Polarization of Protons

High energy spin physics began in early 1960s. It was pointed out that studying the spin dependent forces is one of the most important issues for particle physics. Therefore, it became an urgent need to measure the spin parameters of particle reactions.[1] In order to realize such experiments, it was indispensable to polarize the target protons.

The dynamic method which transfers the electronic polarization to protons using magnetic couplings between electronic and nuclear spins was proposed in France and the US.

Nuclei with magnetic moments in thermal equilibrium at the temperature T and magnetic field B could be oriented to the direction of field. The degree of nuclear polarization P_n with a spin $I = 1/2$ is given by

$$P_n = \tanh(\mu B/kT),$$

where B and μ are the external magnetic field and the nuclear magnetic moment, respectively. It is

$$P_n = \tanh(1.02 \times 10^{-7} B/T)$$

for protons. In principle, the proton polarization of 76% could be obtained at 0.01K and in 10T. However, it is quite difficult to realize such high polarization. On the other hand, we can get sizable polarizations of nuclei in paramagnetic materials by means of the dynamical method in rather lower field and at higher temperature. In the dynamical method magnetic coupling between electron spins and nuclear spins is used to transfer the polarization of electrons to nuclei. The method was proposed by Abragam for paramagnetic materials.[2] The phenomenon was named 'solid effect'.

The energy level diagram of a paramagnetic center coupled to a single neighboring proton in high external magnetic field is shown in Fig. 1(a). The transition a in Fig. 1(a) is a 'forbidden transition' in which both S_z and I_z reverse the sign. While the transition b in which S_z reverses and I_z remains the same are called 'allowed transition'. The dipole-dipole coupling does ensure that there is a small admixture of $S_z = +1/2$ and $-1/2$. The admixture makes forbidden transitions possible by RF field which is applied at the frequency corresponding to the transition a.

The populations of the two levels connected by the forbidden transition can be equalized. It corresponds to the "flip-flop" of an electron spin and a proton spin as shown in Fig. 1(b). Because of the strong coupling between electrons and lattice the relaxation time of electron (T_{1s}) is short, whereas, the proton spin stays on the level for long time since coupling between the proton spin and the lattice is weak. As two neighboring protons are coupled by the dipole-dipole interaction, the orientation of proton spin diffuses throughout whole material, trending to equalize the nuclear polarization.

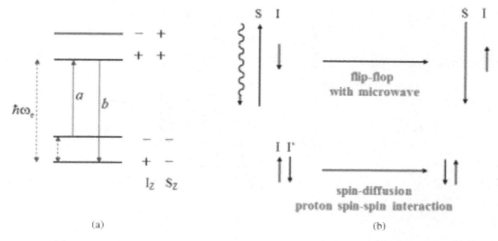

Fig. 1. (a) Energy level diagram of the paramagnetic center coupled to the proton. (b) Flip-flop of an electron spin and a proton spin by means of microwave irradiation. The nuclear spin diffusion is also shown.

At the beginning of the history of dynamically polarized target, protons in the crystal of $La_2Mg_3(NO_3)_{12}24H_2O$, (LMN), containing a small percent (e.g. 0.2%) of Nd and Ce, were polarized by Abragam[3] and Jeffries[4], respectively.

Since the NMR signals for positive and negative polarizations are well-resolved for LMN, the solid effect works so well. The proton polarization of about 80 % was obtained in the temperature around 1 K and the magnetic field of 1.8 T. The LMN targets were successfully operated for elastic scattering experiments with π, K, p and n beams.

3. High Energy Physics with Polarized Target in 1960s

The success of the dynamic polarization was amazing-event for high energy physicists, since it promised a new field of particle physics. At the International Conference on Polarized Targets held at Saclay in 1966, Dalitz pointed out that the polarized target may lead to especially illuminating information on three major areas in particle physics.

(i) High energy scattering where Regge-pole exchange is dominant.

(ii) Tests of time-reversal invariance for electromagnetic processes.

(iii) Hadron spectroscopy. Many resonant states had been observed for mesonic and baryonic states. In the attempt to classify and understand these hadronic states, the first need is for the determination of the spin and the parity for each state.

The first experiment with the polarized target was performed to measure the parameter C_{nn} for p-p scattering at 20 MeV by Abragam, Borghini, Catillon,

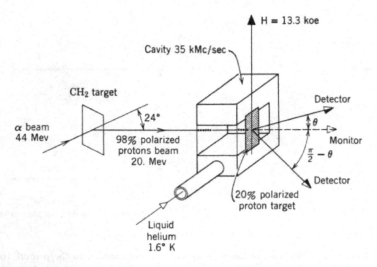

Fig. 2.   Apparatus of a p-p scattering experiment with a polarized target at Saclay.

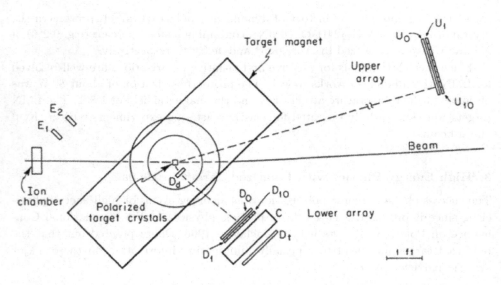

Fig. 3.   Experimental setup for π-p scattering at Berkeley.

Coustham, Roubeau and Thirion at Saclay in 1962[5]. The experiment was done with a polarized beam on a polarized protons in LMN.

    The first pion-proton scattering experiment was performed at Berkeley (Bevatron) by Chamberlain, Jeffries, Schutz, Shapiro, and van Rossum in 1963.[6] In this experiment it was necessary to measure the both angles of pion and proton in order to check the coplanarity, since the background from complex nuclei were enormous.

10GeV/c 12GeV/c

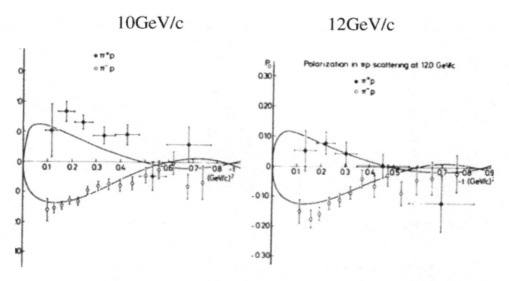

Fig. 4.   Asymmetries of pion-proton elastic scattering measured at CERN.

In 1965, a pion-proton scattering experiment was made with $\sim 1\,\text{GeV}$ pions by Atkinson, Cox, Duke, Heard, Jones, Kemp, Murphy, Prentice and Thresher at Rutherford.[7] It indicated three nucleon resonances, that is, $N^*(1674)5/2$, $N^*(1688)5/2^+$, $N^*(1920)7/2^+$, which had not been previously seen.

A pion-proton scattering experiment was also made with $\sim 2\,\text{GeV}$ pions by Suwa, Yokosawa Booth, Esterling and Hill at Argonne in 1965.[8] The result showed a higher nucleon resonance, $N^*(2190)\,7/2-$.

Asymmetries of pion-proton elastic scattering were measured at 10-12GeV/c by Dick, Borghini et al. at CERN.[1] The results showed a mirror symmetry for π^-p and π^+p scatterings.

The LMN targets were successfully operated for these experiments. And the polarized target became a very important tool for particle physics.

In the mid-1960s LMN polarized targets were constructed in many high energy laboratories, including Berkeley, Argonne, Los Alamos and Harvard in the USA, Saclay and Grenoble in France, Rutherford and Liverpool in UK, Nagoya-INS in Japan, Dubna and Protvino in USSR, and CERN. No polarized target was constructed in SLAC and DESY, because radiation damages of LMN crystal with electrons and γ-beams were serious.

4. Organic Materials for Polarized Target

Although LMN was an excellent polarized target material for elastic scattering experiments, it was not convenient for other experiments, e.g. backward scattering, inelastic scattering, rare decay etc. because of the enormous amount of background

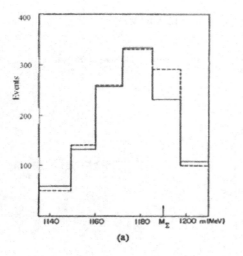

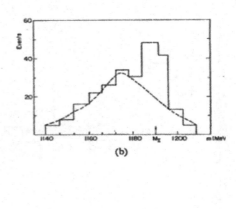

Fig. 5. Missing mass spectra of $\sum$ with (a) LMN and (b) CH_2targets[9].

events related to nuclei different from hydrogen, as the dilution factor (polarizable nucleons/total number of nucleons) is too small. Asymmetry measurements of $\pi^+ +$ $p \rightarrow K^+ + \Sigma^+$ reaction were tried to carry out at CERN and Berkeley in order to test the parity conservation in strong interaction using LMN targets, it was difficult to identify the events from free protons, since the ratio of free protons to bound protons was 1 : 15.

Furthermore, it was difficult to use LMN targets for electron and photon beams, since crystal of LMN is damaged seriously with relativistic particles of $2 \times 10^{12}/cm^2$. Therefore, the dynamic polarization of protons in organic materials was tested with free radicals in the last half of 1960s, since these materials have higher concentration of free protons, and are 250 times stronger for radiation damage. More than 200 kinds of organic materials with several sorts of free radicals had been tested to polarize by Borghini et al. at CERN. Despite the tremendous efforts had been made, none of materials had been successfully polarized up to 25 % before 1968.

Breakthroughs were achieved dramatically in 1969. Protons in butanol with small amount of water doped with porphyrexide were polarized up to 40 % at 1 K and in 2.5 T by Mango, Runolfsson and Borghini at CERN.[10] At the same time protons in diol with Cr^{5+}complex were polarized up to 45 % at 1 K in 2.5 T by Glättli et al. at Saclay.[11] A few months later protons in diol were found to be polarized up to 80 % at the temperature lower than 0.5 K by Masaike et al. at Saclay,[12] while Hill et al. at Argonne succeeded in polarizing protons in butanol up to 67 % in a ^{3}He cryostat.[13] These values are surprisingly large, since it had been believed before their trials that the polarization at lower temperature than 1 K would be less than that at 1 K. Then, polarized targets with diols and butanols cooled in ^{3}He cryostats have been used in most of high energy physics laboratories.

5. Spin Temperature Theory

Abragam and Redfield proposed the spin temperature theory in which a spin system isolated from the lattice and subjected to spin-spin interactions proceeds toward an equilibrium such that the probabilities of finding the system on the energy levels are given by a Boltzmann distribution, $\exp(-E_i/kT_s)$, where T is the spin temperature of the state.[14] In this hypothesis, spin systems behave in the same way as the systems considered in thermodynamics.

The spin temperature theory was applied to dynamic nuclear polarization by Solomon (Saclay) in 1962 following a work of Goldman and Landesman on two nuclear species.[15] They pointed out that the electron non-Zeeman reservoir is in close thermal contact with the nuclear Zeeman reservoir through thermal mixing, and their common spin temperature evolves towards high nuclear polarization.

Borghini reinvented the spin temperature theory for dynamic nuclear polarization in 1968[16] and introduced it to particle physicists at Berkeley Conference on Polarized Target in 1971.[9] He showed a model of spin packets for the inhomogeneous broadening of the electronic resonance lines, with which he computed the nuclear polarization under microwave irradiation in low temperature and named the spin temperature model the 'Donkey Effect'.

Borghini claimed that more than two electrons participate to the dynamic polarization in this model.

6. ^{3}He-^{4}He Dilution Refrigerators for Spin Frozen Target

In 1965 Schmugge and Jeffries discussed the possibility of maintaining the polarization without microwave irradiation, if the nuclear spin relaxation time is long enough.[17] Such a target is advantageous because of the large access angle around

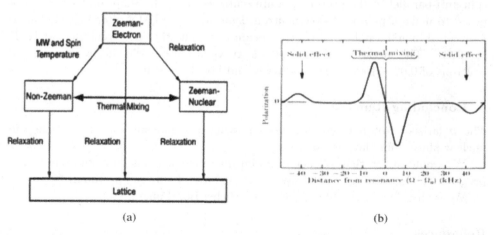

(a)   (b)

Fig. 6.   (a) Thermal couplings between 4 thermal reservoirs: electronic Zeeman, electronic non-Zeeman, nuclear Zeeman and lattice in the course of DNP under microwave irradiation[15]. (b) NMR signals enhanced by sold effect and thermal mixing[14].

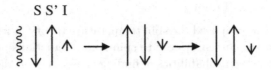

Fig. 7.   Cooling of the nuclear spin by the electron spin-spin interaction reservoir.[9]

the target area in less homogeneous and lower magnetic field than in the polarizing condition.

^{3}He-^{4}He dilution refrigerators were developed toward higher polarization in lower temperature than 0.1K. Neganov (Dubna) and Varoquaux (Orsay) pointed out the possibilities to use the ^{3}He-^{4}He dilution refrigerator for the polarized target in 1966.[1]

Nagamine (Tokyo) made a dilution refrigerator for static polarization of Bi nuclei in 1969.[9]

The first spin frozen target with a horizontal dilution refrigerator was realized by Niinikoski in 1971 and was operated successfully at CERN in 1974.[18]

Soon after the success at CERN, a spin frozen deuteron target was constructed at KEK for the reactions $K^+n \rightarrow K^+n$, K^0p.[19] Then, Saclay group constructed a spin frozen target in late 1970s for nucleon-nucleon scattering experiments.[20] It consisted of a high power dilution refrigerator for target material of 100 cm^3, a vertical polarizing magnet and two holding magnets. Spin frozen targets were also constructed in Bonn, PSI, Dubna and other laboratories.

7. Neutron Polarization with Proton Filter

Neutron transmitted through polarized protons are polarized, since neutrons with spin anti-parallel to the proton spin are scattered away. L. Shapiro at Dubna proposed to make a polarized slow neutron beam using an LMN filter in 1966.[1] Ishimoto et al. made a polarized neutron beam using an ethylene glycol filter at KEK in 1976.[21] The method was used for parity violation experiments with polarized neutrons of $0.02 \sim 1$ eV at Dubna, KEK and Los Alamos in 1980s.

8. Concluding Remarks

The polarized proton target has been playing a significant role in particle and nuclear physics for the last 50 years.

We would stress that Michel Borghini made tremendous contributions to the progress of 'High Energy Spin Physics' through polarizing the proton in 1960s.

We wish to honor the life and work of Michel Borghini.

References

1. in *Proceedings of the International Conference on Polarized Targets and Ion Sources*, Saclay, France 1966 (ed. by CEN-Saclay).

2. A. Abragam, *Phys. Rev.* **98**, 1729 (1955).
3. A. Abragam et al. *Phys. Lett.* **2**, 310 (1963).
4. T. J. Schmugge, C. D. Jeffries, *Phys. Rev. Lett,* **9**, 268 (1963).
5. A. Abragam et al. *Phys. Lett.* **2,** 310 (1962).
6. O. Chamberlain et al. *Phys. Lett.* **7**, 293 (1963).
7. H. H. Aktinson et al. *Proc. Roy. Soc.* **A289**, 449 (1966).
8. S. Suwa et al.*Phys. Rev. Lett.* **15** 560 (1965).
9. in *Proceedings of the 2nd International Conferece on Polarized Targets* (Berkeley 1971).
10. S. Mango, O. Runolfsson and M. Borghini, *Nucl. Instr. Meth.* **72**, 45 (1969).
11. H. Glättli et al. *Phys. Lett.* **29A**, 250 (1969).
12. A. Masaike et al. *Phys. Lett.* **30A**, 63 (1969).
13. D. Hill et al. *Phys. Rev. Lett.* **23**, 460 (1969).
14. M. Goldman, *Spin Temperature and Nuclear Magnetic Resonance in Solid* (Oxford University Press 1970).
15. I. Solomon, in XI *Colloque Ampere on Magnetic and Electric Resonance and Relaxation,* p.25 (1962) Eindhoven, Netherland (North-Holland, Amsterdam).
16. M. Goldman and A. Landesman, *Phys. Rev.* **132**, 610 (1963).
17. M. Borghini, *Phys. Rev. Lett.* **20,** 419 (1968).
18. T. J. Schmugge, C. D. Jeffries,*Phys. Rev.* **138A**, 1785 (1965).
19. T. Niinikoski, F. Udo, *Nucl. Instr. Meth.* **134**, 219 (1976).
20. S. Isagawa et al. *Nucl. Instr. Meth.* **154**, 213 (1978).
21. J. Deregel et al. in*Proc. Int. Conf. on High Energy Physics with Pol. Beam and Targets,* p. 463 (1980) Lausanne.
22. S. Ishimoto et al. *Japanese Jour. Appl. Phys.* **25** 1246 (1986).

Spin Physics (SPIN2014)
International Journal of Modern Physics: Conference Series
Vol. 40 (2016) 1660115 (10 pages)
© The Author(s)
DOI: 10.1142/S2010194516601150

World Scientific
www.worldscientific.com

Today's Polarized Solid Targets in Borghini's Footsteps

Werner Meyer

Ruhr-Universität Bochum, EP1 AG1, Universitätsstr. 150,
44780 Bochum, Germany
meyer@ep1.rub.de

Published 29 February 2016

The development, in the early 1960's, of the dynamic nuclear polarization process in solid diamagnetic materials, doped with paramagnetic radicals. led to the use of solid polarized targets in numerous nuclear and particle physics experiments.

In 1965 Michel Borghini moved from Saclay to CERN, where under his leadership of the polarized target group (until 1978) decisive developments have been made. Target material research and the early frozen spin targets were certainly the basis for the successful story of present day polarized solid targets.

PACS numbers: 29.90.+r

1. Introduction

The prehistory of the Dynamic Nuclear Polarization (DNP) process can be traced back to the late 1940s, specifically to the discovery of the Nuclear Magnetic Resonance (NMR), the simultaneous development of the Electron Paramagnetic Resonance (EPR) and the elucidation of nuclear and electronic spin lattice relaxation processes in solids. These events were, of course, interwoven with the broad and rich development of solid state physics. In 1953 Overhauser made the brilliant prediction, that, for a metal, the radio frequency (RF) saturation of the ESR line of the conduction electrons would enhance the nuclear polarization by a large factor μ_e/μ_n. The next step was taken by Abragam to extend the Overhauser Effect to nonmetals. The first published data for a clear DNP process have been made in 1959. It took another 3 years that large polarizations were achieved by DNP: 70% obtained at 1.5 K and 3.6 T. The first polarized solid target for particle physics experiments has been constructed at Saclay in 1962 (A. Abragam *et al.*), followed by the first polarized solid-target for "high energy" physics experiments at Berkeley (0. Chamberlain *et al.*).

Since that time, various polarized solid targets have been constructed and have been operated in numerous particle physics experiments worldwide. The quality of such targets is measured in terms of their content of polarizable nuclei compared to all nuclei in the target material and of course depends on the degree of the polarization. The 1970s and early 1980s can be considered as the bloom of the development of these targets, where important technical developments have been employed (i.e. ^{3}He/^{4}He dilution refrigerators and superconducting magnets), to enhance the polarization degree in more appropriate target materials. At that time contacts to chemists were a normal process. Decisive for the progress was the fact that during this time experimental developments have been accompanied by theoreticians and their work, too. The theory which correctly describes the behavior of a dipolar coupled spin system under saturation is the Provotorov theory. Within this framework an expression for the spin temperature can be derived, which depends not only on the degree of saturation but also on the width of the resonance line as well on the time constants of the spin systems. With the exception of the solid state rate equations so far the considerations were restricted to the so called high temperature approximation. Due to the fact that this assumption is far from being valid under usual conditions of a DNP experiment, M. Borghini found a way, which allows the predictions of the spin theory to be tested also in the low temperature regime within special samples. Examples are the partially and fully deuterated alcohols and diols (all of them chemically doped), in which the proton polarization was either compared to the deuteron or to the ^{13}C polarization.

A comprehensive overview about the achievements made in this field up to the late 1980s together with the particular references can be found in Ref. 1. More recent overviews are given in Ref. 2, 3.

2. DNP Parameters

The principles of dynamic nuclear polarization were derived in the 1960s and 1970s[4,5] and have been reviewed frequently, e.g. in Ref. 3. Here, only a phenomenological description of the DNP process and its relevant parameters will be given, since a comprehensive theoretical treatment of the DNP mechanisms is beyond the scope of this article.

Dynamic nuclear polarization works at low temperatures and high magnetic fields by transferring the high electron polarization of a paramagnetic species, which has to be added to the host material by chemical doping or irradiation, to the nuclei with nonzero spin in the host material. This polarization transfer is achieved in terms of microwave irradiation with frequencies close to the electron spin resonance.

Since polarization is understood as the population difference between corresponding Zeeman levels, and since the thermal equilibrium (TE) polarization is described by the Brillouin function of the Boltzmann factor, high magnetic fields and low temperatures are a precondition in order to achieve high polarization values.

In the dynamic case, i.e. under irradiation with microwaves inducing combined electronic and nuclear spin flips, several additional parameters determine the efficiency of the polarization transfer from the paramagnetic electrons to the nuclei, and hence the achievable nuclear polarization. These are quantities like microwave power fed to the sample, cooling power of the polarization refrigerator, the relation of the EPR linewidth to the nuclear Larmor frequency, and nuclear as well as electronic relaxation times, which are dependent on the properties and the concentration of the used radical. Some of these parameters show either a magnetic field dependency (e.g. the EPR linewidth), a temperature dependency or both (e.g. relaxation times). Hence for a given sample there might be an optimum in the DNP conditions, instead of a continuous increase of polarization with increasing magnetic field and decreasing temperature.

In this article special focus is put on recent alcohol target material research (especially for deuterated materials) and recent developments on the use of the so called frozen spin target technology

3. Solid Target Materials Doped with Trityl Radicals

In 1965 one priority of polarized target research at CERN like in many other laboratories was the development of 'better' polarized target materials using chemical dopants for DNP. I recommend for reading to anybody interested in this type of research the article of M. Borghini 'Choice of substances for polarized proton targets'.[6] The breakthrough happened in 1968 with the development of a butanol + 5 % H_2O + porphyrexide mixture.[7]

Since then the paramagnetic radicals of all of the solid target materials used in particle physics experiments feature EPR lines with linewidths at least as large as the corresponding nuclear Larmor frequencies. Thus for a correct description of the DNP process the spin temperature theory is the only valid model.[5] Whereas proton polarization values of more than 90 % could be obtained already in the early 1970s,[8] deuteron polarization values have been considerably lower, typically 30-50 %. This changed with the use of trityl radicals for deuterated compounds. After very promising studies of Ox063 (AH 100136 sodium salt) in propanediol-d_8 and 'Finland 036' (AH 110 355 deutero acid form) in butanol-d_{10} with deuteron polarizations of 80 % at 2.5 T in a dilution refrigerator,[9] a trityl doped butanol d_{10} target was successfully used in the GDH experiment at MAMI.[10] To ensure sufficient solubility of the radical, different types of trityl are used depending on the polarity of the host material. While Ox063 is readily solved in highly polar substances like diols, it is not soluble in the longer chained alcohols as it is the case for 'Finland 036'. Both radicals feature EPR linewidths of about four times smaller than the conventionally used radicals (like Porphyrexide and TEMPO as members of the nitroxyl family) and one order of magnitude smaller than EHBA-Cr(V) at 2.5 T as reported in Ref. 11. In this reference the implication of the EPR linewidth on the maximum nuclear polarization of deuterated solid target materials is discussed in detail.

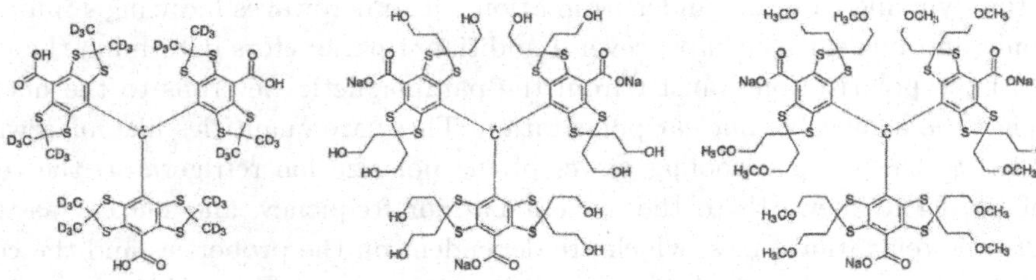

Fig. 1.   Molecular structure of three trityl radicals. Left: 'Finland D36' (AH 110355 deutero acid form) used for butanol-d10 Center: Ox063 (AH 100 136 sodium salt) used for propanediol-ds Right: Ox063Me (AH 111 501 sodium salt) used for pyruvic acid.

Due to their rather small gyromagnetic ratio ^{13}C nuclei are — similar to deuterons - comparably difficult to polarize highly with conventional radicals. Because of their small EPR linewidth, even at higher fields, trityl radicals work very well as DNP agents also for ^{13}C nuclei. For the use with pyruvic acid, which is quite polar, a third type of trityl Ox063Me is used, where the 12 hydroxyl groups are substituted with methoxy groups (see Fig. 1) to prevent reactions of the pyruvic acid with the radical. The EPR linewidth of Ox063Me is slightly larger than that of the Ox063 radical (0.28 mT compared to 0.22 mT at 335 mT).

4. Frozen Spin Targets for Particle Physics Experiments

An important question in any experiment is the optimization of the reaction counting rate

$$N = \mathcal{L}\frac{d\sigma}{d\Omega}\Delta\Omega \tag{1}$$

where $d\sigma/d\Omega$ is the cross section of interest and $\Delta\Omega$ the detector solid angle. The luminosity

$$\mathcal{L} = I \cdot n_t \tag{2}$$

is defined as the product of beam current I (number of beam particles per second) an the areal target thickness n_t (number of target particles per cm^2). Thus, luminosity is one of the main quality factors in determining how fast an experiment can be done.

The quality of a polarization experiment is, however, not only given by the reaction counting rate, but is very much dependent on other factors like background from other sources and the polarization degree of the target material. In designing an experiment, one has to consider seriously these factors in addition to the luminosity. A 'Figure of merit' for polarized solid targets $FOM_{solid\,target}$, can then be defined

$$FOM_{solid\,target} = n_t \cdot P_t^2 \cdot f^2 \tag{3}$$

i.e. the larger $n_t \cdot P_t^2 \cdot f^2$ the shorter the running time required to achieve a chosen accuracy in the measured asymmetry observed observable. Here P_t is the measured target polarization degree and f is the dilution factor (ration of polarizable nucleons to all nucleons in the target material). As can be seen from eq. (1) another factor is $\Delta\Omega$, which in the optimal case can be 4π.

Depending on the beam intensities there are two modes of operation for a polarized solid state target system. (Polarized) electron beams have to be operated at intensities of less than 100 nA in order to handle cooling requirements and radiation damage of the target materials. In this case polarized targets with ^{4}He evaporative cooling are used, which have to work in a continuous mode, i.e. with permanent microwave irradiation to maintain the DNP process at 1 K.[12] At this temperature the nucleus (nucleon) polarization relaxation time T_1 is relatively short. This continuous mode operation puts strong constraints on the design of the polarizing magnets. Due to the field homogeneity requirements over the entire target volume, the large dimensions of the magnet coils limit the angles for the outgoing particles. This mode of operation has been used for the SLAC experiments in the 1990s and is presently used for experiments at TJNAF (see Fig. 2 (l)). Other electromagnetic probes such as tagged photon beams are intensity limited by their production techniques to about 10^8 particles/s. To obtain a reasonable counting rate, a wide opening angle with the ability to simultaneously measure a large kinematic range is needed. This can be achieved with the concept of the frozen spin target.

The operation of the frozen spin target is based on the experimental fact that the nucleon polarization relaxation time T_1 is a very strong function of the temperature and magnetic field. T_1 characterizes the polarization decay after switching off the DNP mechanism (microwaves). Typical values for T_1 are minutes at a temperature of 1 K and days below 100 mK. The principle of the frozen spin target operation is to polarize the target material at a high field (e.g. 2.5 T) and in the temperature range 0.15–0.3 K. Once the target material is optimally polarized, the microwaves

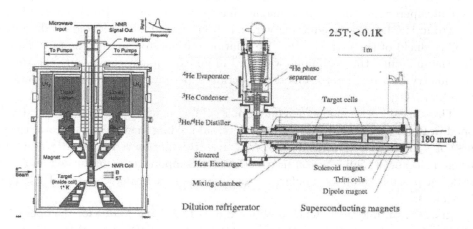

Fig. 2. (l) 5 T polarizing superconducting magnet with a vertically operated ^{4}He evaporation refrigerator. (r) The CERN dilution ^{3}He/^{4}He dilution refrigerator with a superconducting 2.5 T magnet and a 0.5 T dipole magnet.

are switched off and the polarization is frozen at temperatures around 50 mK (frozen spin mode). Due to the very long T_1 at these temperatures, it is possible to reduce the field to a value of around 0.5 T where the polarization decay is acceptable (holding field).

A very specific target design has been developed for the CERN polarized target experiments SMC and COMPASS.[13] A huge ^{3}He/^{4}He dilution refrigerator (T $\sim$ 50 mK) is connected to a 2.5 T superconducting solenoid which is surrounded by a dipole magnet (see Fig. 2 (r)) . With a beam intensity of $\sim 10^7 \mu^+/s$, which was more than four orders of magnitude less than at the SLAC experiments, the luminosity was increased by having a target two orders of magnitude longer than at SLAC or TJNAF. In the CERN experiments it was not possible to flip the beam polarization (unless one changed to μ^- !) and with a polarization build-up time of hours, it was not very easy to change the target polarization frequently. So a strategy of rotating the solenoid field to the opposite direction was adopted. With 190 GeV muons and a longitudinal field, the effect on the scattered particles was minimal. A dipole magnet was turned on at a strategic time so that the total magnetic field never went below 0.5 T while the solenoid was crossing zero and the target was in the frozen spin mode. Thus the CERN polarized target can be operated in the continuous mode at 2.5 T as well as in a frozen spin mode $T \leq 50$ mK for horizontal or transverse polarization direction. Other systems using the frozen spin mode with external holding magnets have been operated by SACLAY[14] and PSI[15] groups.

The disadvantage of the external holding coil arrangement is the large size and strong fringe field of the superconducting magnets. Therefore such frozen spin targets cannot be operated in combination with detection systems, where modern detector components are placed close to the target refrigerator. To overcome this problem, the scheme of an internal superconducting 'holding coil' has been discussed in 1980 by Niinikoski[16] but not technically realized. In 1992 a new type of a small superconducting holding magnet has been built by the Bonn polarized target group.[17] The superconducting wire has been wound as a solenoid on the inner cooling shield of the vertical ^{3}He/^{4}He-dilution refrigerator around the target area. For the first time this technique was used in an experiment to measure the target asymmetry in a real photon photoproduction experiment at the Bonn accelerator ELSA in 1994.

This concept of a small superconducting holding magnet has been further developed in Bonn (see Fig. 3) leading to an internal holding magnet with a total thickness of $\sim$1 mm, placed inside a horizontal ^{3}He/^{4}He dilution refrigerator.[17] Thus the fringe field and its influence on nearby detector components is minimized and that makes it possible to operate polarized solid targets in a 4π-detector (e.g. DAPHNE in Mainz).[18]

In the meantime this scheme is adopted by TJNAF and Mainz for the operation of their frozen spin targets in the CLAS detector and CRYSTAL BALL detector, respectively.

Fig. 3. The Bonn superconducting holding magnet (0.6 T) with a total thickness of ~ 1 mm, placed inside a horizontal ^{3}He/^{4}He dilution refrigerator.

Meanwhile the Bonn, Mainz and TJNAF polarized target groups are able to replace the internal solenoidal holding coil by a saddle coil geometry which gives access to y- and x-polarization directions in the frozen spin mode.

5. Perspectives

The frozen spin target operation is a powerful tool for (double) polarization experiments at low intensity beams. Nevertheless the price for the good particle detection acceptance is a loss in beam time, which is needed to repolarize the target material. Furthermore a dedicated railway system to move the detector and external polarizing magnet for the polarization procedure is needed. In best cases a beam time efficiency of 80 % is attainable. The second restriction from the 'figure of merit' point of view is the decay of the polarization during the data taking. The overall target polarization of an experiment using a frozen spin target is given by the nucleon relaxation time and limited to approximately $0.8\ P_{max}$. The goal for the future is to combine the advantages of the frozen spin technique with those of the 'continuous mode' operating target to a so called '4π continuous mode' polarized target. This new polarized target scheme leads to an improvement in the ' figure of merit' by a factor of 2 compared to the existing frozen spin target operation. The important points of such a target system are:

- large angular acceptance close to 4π
- high average polarization during the data taking (continuous DNP at high polarizing field and low temperature)
- high luminosities up to $10^{33} cm^{-2} s^{-1}$
- no moving system for the polarization process required
- good beam time efficiency (fast repolarization in case of radiation damage)

In practice, starting from the existing ' internal superconducting holding coil' a new coil capable of providing an increased field (polarizing field) has to be implemented into the ^{3}He/^{4}He refrigerator as an internal polarizing magnet. It has to fulfill the requirements of the high homogeneity of the external polarization magnet and the low mass distribution of the internal holding coil to ensure a good detection probability for the outgoing particles. The most crucial point which has to be considered in the development of the internal polarizing magnet is the homogeneity of the small coil. Because of the minimized dimensions of the solenoid the magnetic volume is only a factor of about 20 larger than the target volume, which needs the good homogeneity of $\Delta B/B \leq 10^{-4}$. The investigation of this problem has led to a new approach for the design of this magnet, which relies on the geometrical shape of the winding.[19]

Presently internal thin superconducting magnets with a field strength of ∼2.0 T and a field homogeneity $\Delta B/B$ of 10^{-3} have been manufactured in Mainz and Bonn[20] and first tests in a ^{3}He/^{4}He dilution refrigerator will follow soon.

As mentioned above, luminosities of up to $10^{33} cm^{-2} s^{-1}$ at high polarization can be achieved with this '4π continuous mode' polarized target i.e. beam intensities higher than $10^{10} particles/s$ upon some extended target area ($\sim cm^2$) can be accepted. However, at such levels of beam intensities the radiation resistance of the polarization due to material damage has to be considered. Then inorganic materials like ammonia (NH_3 and ND_3) or ^{6}LiD and ^{7}LiH are the target materials of choice.[21–23]

6. DNP Polarized Target Materials for Medical Applications

Magnetic Resonance Imaging (MRI) is a very successful imaging modality in terms of temporal and spatial resolution of morphological details. The sensitivity of NMR, imaging or spectroscopy, is strongly limited by the low polarization that can be reached at room temperature and any achievable field strength. In an attempt to improve this, DNP-hyperpolarization was invented. Several imaging applications will benefit from the strong signal available from the ^{13}C agent, e.g. angiography and perfusion studies. Other applications have so far been impossible by MRI, e.g. studying the metabolic fate of a hyperpolarized agent during the first minute after injection. The ability to study metabolism of endogenous molecules could have great implications in e.g. oncology and cardiology. Conventional MRI contrast parameters have not been correlated with processes at the cellular level. The idea behind hyperpolarized ^{13}C — labeled contrast agents was to directly image by MRI an agent (e.g. endogenous molecule) that would have specificity to certain deceases.

An apparatus and method for producing strongly polarized molecules in solution has been developed.[24] The nuclear spins of a solid sample doped with a paramagnetic agent (Trityl or Tempo radicals) can be strongly polarized at low temperature (∼ 1.2 K) and moderate magnetic field (3.35 T) by exposing the sample to microwave irradiation (94 GHz) close to the electron paramagnetic resonance. The solid sample is rapidly dissolved and an image or spectrum is acquired within a time of the order of the nuclear relaxation time T_1 (within minutes). By this method it has been possible to produce solutions of ^{13}C — labeled molecules with polarization of up to ∼ 40 %.[25] A ^{13}C-polarization of ∼80 % has been measured in trityl doped pyruvic acid by increasing the magnetic field up to 5.0 T.[26] This very high ^{13}C polarization is certainly a result of the small EPR — linewidth of the trityl radical even at high field.[11] As it is shown elsewhere[27] the DNP mechanism in trityl doped partially deuterated pyruvic acid is Borghini's spin — temperature model.

Progress in this field is reported in a series of DNP symposia, the latest held in Copenhagen (2013).

7. Conclusion

The subsequent developments of all aspects of a polarized solid state target system have opened up new experimental possibilities and allowed a choice of approaches to a particular physics problem. There are a variety of polarizable solid target materials available with dilution factors varying between 0.13 and 0.5. The highest polarization resistance against radiation damage is shown in materials which are prepared for DNP by irradiation. Polarizing them in ^{4}He evaporation refrigerators with their high cooling power at 1 K beam intensities upon the targets up to $\sim 6 \cdot 10^{11}$ particles/s can be tolerated. Thus luminosities of higher than $10^{35} cm^{-2} \cdot s^{-1}$ can be achieved.

Polarized target experiments performed with low intensity beams such as polarized muons or (polarized) tagged photons have to rely on solid target materials. At intensities up to 10^8 particles/s the frozen spin mode operation of the target system is favorable in order to allow a particle detection over a large (4π) angular acceptance.

Maximum polarization values for protons $> 90\%$ and for deuterons up to 80 % are achieved. Polarization measurements are done by the NMR technique to a relative accuracy of $< \pm 3$ % for protons and $< \pm 5$ % for deuterons. Future developments as discussed will certainly improve the experimental situation for experiments with polarized solid targets in the next decade.

Latest developments open up the possibility to use DNP-hyperpolarized substances in the field of medicine and others. Thus DNP experiences a second springtime and a new huge research community will profit from it enormously.

Finally I conclude with a remark of the Nobel laureate O. Chamberlain, who listed in 1985 the people that he felt to be "the heroes — the people who have given spin physics a special push". M. Borghini is the only one that he cites twice: with A. Abragam and colleagues for the first polarized target and the first experiment to use such a target; and with T. Niininkoski, for their introduction of the frozen spin target and showing the advantages of powerful (dilution) refrigerators.

For me his name on its own is manifested in DNP theory by the developments of the so called 'Borghini spin temperature model'.

References

1. C.D. Jeffries, in: K.H. Althoff, W. Meyer (Eds.), *Proceedings of the 9th International Symposium High Energy Spin Physics*, (2000) Vol.1 pp. 3-19, and references therein.

2. D.G. Crabb, W. Meyer, *Annu. Rev. Nucl. Sci. 47*, (1997) 67.
3. S. Goertz, W. Meyer, G. Reicherz, *Prog. Nucl. Phys. 49* (2002) 403.
4. A. Abragam, *Principles of Nuclear Magnetism*, Oxford University Press (1961).
5. M. Goldman, *Spin Temperature and Nuclear Magnetic Resonance in Solids*, Oxford University Press (1970).
6. M. Borghini, *CERN Report 66-3* (1966).
7. S. Mango *et al.*, *Nucl. Instr. and Meth. 72* (1969) 45.
8. W. de Boer, T.O. Niinikoski, *Nucl. Intr. and Meth.* 114 (1974) 495.
9. S. Goertz, J. Harmsen, J. Heckmann, C. Hess, W. Meyer, E. Radtke, G. Reicherz, *Nucl. Instr. and Meth. A 526* (2004) 43.
10. A. Thomas, *Eur. Phys. J. A* 28(1) (2006) 161.
11. J. Heckmann, S. Goertz, W. Meyer, E. Radtke, G. Reicherz, *Phys. Rev. B 74* (2006)
12. D.G. Crabb *et al.*, *Nucl. Instr. and Meth. 356* (1995) 9
13. J. Kyynärännen *et al.*, *Nucl. Instr. and Meth. 356* (1995) 47.
14. R. Bernard *et al.*, *Nucl. Instr. and Meth. 249* (1986) 176.
15. B. van den Brandt *et al.*, *Nucl. Instr. and Meth. 356* (1995) 53
16. T.O. Niinikoski, *Int. Conf. on High Energy Physics with Pol. Beams and Pol. Targets*, Lausanne (1980) pp. 191
17. H. Dutz *et al.*, *Nucl. Instr. Meth. A 356* (1995) 111
18. H.-J. Arends *et al.*, *Phys. Rev. Lett. 84* (2000) 5950
19. A. Raccanelli, H. Dutz, R. Krause, *Proc. of the 11th Workshop on Pol. Sources and Targets*, Tokyo (2005) World Scientific (2007) 221
20. H. Dutz and A. Thomas, *private communications*
21. W. Meyer, *Nucl. Instr. and Meth. A 526* (2004) 12
22. D.G. Crabb, *these Proceedings*
23. J. Ball, *Nucl. Instr. Meth. A 526* (2004) 7
24. J. Ardenkjaer-Larson *et al.*, *Proc. Math. Acad. Sc.* USA (2003) 10158
25. J. Wolber *et al.*, *Nucl. Instr. and Meth. A526* (2004) 173
26. W. Meyer *et al.*, *Nucl. Instr. and Meth. A 631* (2011) 1
27. F. Greffrath, Diploma Thesis, Ruhr-Universität Bochum, Germany (2008) (available from the institutes webpage: www.ep1.rub.de/poltarg/hompage_files/theses.html)

Spin Physics (SPIN2014)
International Journal of Modern Physics: Conference Series
Vol. 40 (2016) 1660116 (10 pages)
© The Author(s)
DOI: 10.1142/S2010194516601162

Michel Borghini as a Mentor and Father of the Theory of Polarization in Polarized Targets

Wim de Boer

Karlsruhe Institute of Technology, Physics Department, IEKP, Campus Süd,
Postfach 6980, 76049 Karlsruhe, Germany
wim.de.boer@kit.edu

Published 29 February 2016

This paper is a contribution to the memorial session for Michel Borghini at the Spin 2014 conference in Bejing, honoring his pivotal role for the development of polarized targets in high energy physics. Borghini proposed for the first time the correct mechanism for dynamic polarization in polarized targets using organic materials doped with free radicals. In these amorphous materials the spin levels are broadened by spin-spin interactions and g-factor anisotropy, which allows a high dynamic polarization of nuclei by cooling of the spin-spin interaction reservoir. In this contribution I summarize the experimental evidence for this mechanism. These pertinent experiments were done at CERN in the years 1971 - 1974, when I was a graduate student under the guidance of Michel Borghini. I finish by shortly describing how Borghini's spin temperature theory is now applied in cancer therapy.

Keywords: Dynamc polarization; spin temperature theory; cancer therapy.

1. Introduction

Studying the effect of spin in particle interactions has been a topic of interest in high energy physics (see e.g. Ref.[1] for a review), which required polarized particles, either as target or as beam or both. The development of polarized targets at CERN was driven by Michel Borghini, while Alan Krisch from the University of Michigan (Ann Arbor) pushed the polarized beams at Argonne and later at Brookhaven and other accelerators.[2] I was lucky enough to work with both of them. After finishing my Master thesis at the Technical University of Delft on studying spin systems in LMN,[3] a material used initially for polarized targets, I came to CERN as a fellow in Michel Borghini's group and contributed heavily to the experiments leading to the acceptance of Borghini's mechanism of dynamic nuclear polarization (DNP) in organic materials[4,5] by "Dynamic Orientation of Nuclei by Cooling of the Electron

Spin-Spin Interactions"[a]. Here the spin-spin interactions (SS) comprise all the non-Zeeman energies, which can broaden the Zeeman levels of the free electrons beyond the nuclear Zeeman levels, thus allowing a thermal contact between the SS-reservoir and the nuclear Zeeman reservoir by electron spin flips in combination with nuclear spin flips. Such a thermal contact is driven at low temperatures mainly by induced spin flips from the polarizing RF field. This dual role of the external RF field (cooling and establishing thermal contact) at low temperatures and in high magnetic fields was the main new idea from Michel, since DNP by cooling of the spin-spin interactions had been demonstrated before in 1963 by Goldman and Landesman[6] in the group of A. Abragam, the world leading expert on DNP at Saclay. but its application to amorphous materials at low temperatures was far from clear.

Borghini, also working in Abragam's group, wrote down his ideas in an extensive thesis. However, his thesis was not accepted by Abragam for reasons unkown to me, but presumably because it lacked experimental verification. Michel's proposed mechanism was clearly proven by our experiments at CERN, done at temperatures down to 0.1K and magnetic fields up to 5 T. After all our results were published by 1976,[7-13] Abragam and Goldman wrote a review on DNP, describing in detail our results and recognizing that this was the mechanism of DNP in polarized targets.[14] These papers were the basis of my PhD thesis[15] at the Technical University of Delft. Promotor was Prof. B.S. Blaisse and Michel was a member of the thesis committee, as shown in Fig. 1.

When I came to Borghini's group in 1971, scattering experiments with polarized butanol targets[16] were in full swing. However, a higher proton polarization was requested and we continued to work on Michel's list of possible materials, which should be tried. This was extremely tedious, since every material could be doped with every free radical in a range of concentrations. Michel left every one much freedom in trying out ideas and organizing his work. This fosters the creativity of the individual much stronger than in a hierarchical group structure, where everyone is told what to do. I have kept this working style in my working groups. We rarely had group meetings, but Michel regularly informed himself how things were going and took care that the infrastructure was optimal, so we had an outstanding mechanical workshop with Jean-Michel Rieubland as driving force behind the actual building and running of the polarized targets, George Gattone as head of the chemistry laboratory, Fred Udo and Huib Ponssen, also two dutch staff members, providing the electronics and digital readout of the polarization. It included for every target an HP2100 computer, which allowed not only a precision determination of the polarization by averaging the sometimes tiny thermal equilibrium signals, but I could also do all calculations for my thesis on my personal computer. The programs were punched with a Teletype writer on paper rolls, which in turn could be read by optical readers. All this was high tech at that time. Then there was of

---

[a]Michel called this the "DONKEY" effect, but the name did not stick.

Fig. 1. a) My thesis Committee at the Technical University of Delft (1974) with Michel on the right. b) Discussion with Michel after the thesis exam.

course Tapio Niinikoski, the cryogenic genius, who obtained his thesis on the development of the horizontal dilution refrigerators in the group of Prof. O.V. Lounasma at the Helsinki University of Technology, roughly at the same time as I received my Ph.D thesis at the Technical University of Delft. Tapio was the mastermind behind the frozen spin targets[17] and became the head of the polarized target group after Michel took over other responsibilities at CERN. After getting to know all the tricky details of how to build and operate a polarized target, I was hired by Alan Krisch at the University of Michigan in Ann Arbor, who had just started to do experiments with polarized beams at Argonne National Laboratory and installed a polarized target. This enabled us to measure the cross section for a polarized proton beam on a polarized proton target. Surprisingly, this led to significant spin effects, even in the total elastic pp cross section for protons with spins parallel or antiparallel, an experimental result,[18] which still lacks an interpretation in the frame work of QCD. But for me the main result was, that high energy physics is at least as interesting as solid state physics, so I became a particle physicist[b]. In this contribution I want to summarize the exciting experiments done between 1971 and 1974 in Borghini's group at CERN.

2. The Theory of Dynamic Polarization

In solid *crystalline* materials doped with a small concentration of paramagnetic centers with an unpaired free electron the mechanism of dynamic polarization is easy to understand: in a magnetic field H at a temperature T the relative fraction of the free electrons n_i over the spin states with energy E_i is given by the Boltzmann distribution $n_i = exp(-E_i/kT_S)$ for a spin temperature T_S. This leads for a spin $1/2$ to a polarization $P = (n_+ - n_-)/(n_- + n_-) = \tanh(h\nu/2kT_S)$, where h is Planck's constant and ν the Larmor frequency of the spin system. At high temperatures one can expand the exponential expressions, in which case the electron and nuclear

---

[b]The possible unification of all forces in Supersymmetry[19] spurred my interest in dark matter,[20] so I later joined, in addition, the astroparticle physics community in search for the elusive dark matter.

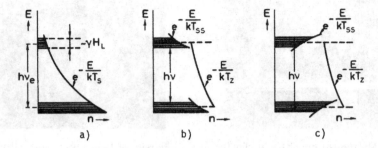

Fig. 2. a) The equilibrium spin temperature T_S determines the population of the energy levels in a magnetic field. Off-resonance microwave radiation below (above) the Larmor frequency populates the lower (upper) levels inside the broadened Zeeman levels, thus cooling (heating) the SS-reservoir, as indicated in b) (c). This leads to different temperatures of the Zeeman - and SS-reservoir, indicated by T_Z and $T_S S$, respectively.

polarization are related to the inverse spin temperature $\beta = h/kT_S$ by $P_e = -\beta\nu_e/2$ and $P_n = \beta\nu_n/2$. Numerical results for the Larmor frequency and a spin temperature T_S equal to the lattice temperature lead to an electron polarization $P_e = -0.9975$ in a magnetic field of 2.5 T and a temperature T of 0.5 K, while the proton polarization $P_n = +0.00511$ under the same conditions. DNP is the art of transferring the high electron polarization to the nuclei via microwave induced spin flips.

For *crystalline* materials the dominant DNP mechanism is the "solid" effect (also called solid-state effect), which was proposed by Abragam and Proctor and verified experimentally, see the review[14] for original references. In this case one stimulates by microwave irradiation the "forbidden" transitions, in which case an electron and neighboring nucleus simultaneously change their spin orientation (either flip-flip or flip-flop transitions, where a flip (flop) indicates a transition to a higher (lower) Zeeman level). The electron will return to the ground state quickly with a time constant given by the short electron spin-lattice relaxation time of the order of ms. The nucleus has a much longer spin lattice relaxation time, so it does not quickly return to the ground state, but instead it can transfer its polarization to neighboring nuclei via flip-flop spin transitions. This leads to spin diffusion, which is fast, since energy and angular momentum are conserved. The electron is now ready to polarize the neighboring nucleus again, if it is still receiving photons with the correct energy from an external microwave field. This combination of an external microwave field polarizing neighboring nuclei combined with fast nuclear spin diffusion allows to effectively transfer the high polarization from the electrons to the nuclei.

In *non-crystalline* solids the spin levels are usually broadened by the different orientations of the molecules, which experience different internal magnetic fields and this broading is usually larger than the Zeeman splitting of the nuclei. In this case the resolved solid-state effect will not work anymore, since one is stimulating

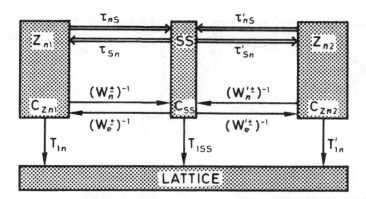

Fig. 3.   Schematic diagram of the thermal contact between two nuclear Zeeman reservoirs Z_{ni} with heat capacities C_{Zni}, the spin-spin interaction reservoir SS and the lattice. The double arrows indicate the thermal contact via the flip-flop transitions between two electrons accompanied with a nuclear spin flip, while the single arrows indicate the microwave induced forbidden transitions of a simultaneous electron and nuclear spin transition.

simultaneously flip-flop and flip-flip transitions[c]. However, another mechanism of DNP may become effective, which is most easily explained by first introducing the concept of a spin temperature and a spin-spin interaction (SS) reservoir.[6, 21] These concepts are visualized in Fig. 2. In a) the populations of the energy levels follow the Boltzmann distribution, both, for the large Zeeman splitting of the electrons and inside an energy band for a given Zeeman level. However, inside a band, whose width is determined by the non-Zeeman interactions, like the g-factor anisotropy or spin-spin interactions (SS), the population can be changed by external photons, if one irradiates with microwave frequencies slightly different from the central Zeeman frequency. This can either cool (Fig. 2b) or heat (Fig. 2c) the SS-reservoir[22, 23] and lead even to the highest levels having the highest population, as shown in Fig. 2c, which corresponds to negative spin temperatures of the SS-reservoir. The question is: how strong is the thermal contact between the nuclear Zeeman energy reservoir and this SS-reservoir? This contact can be established either by (i) spontaneous electron spin flip-flops between the Zeeman levels with a simultaneous nuclear spin flip (so a 3-spin process) or this contact can be established by (ii) the microwave induced forbidden transitions of the "solid" effect. The different transitions for the thermal contact are schematically indicated in Fig. 3. The thermal contact via (i) was demonstrated by Goldman and Landesman,[6] who first cooled the SS-reservoir by off-resonance RF irradiation. They then switched on a magnetic field, which revealed a nuclear polarization, obviously obtained from the thermal contact with

---

[c]The net polarization is then given by the difference in intensity of the flip-flop and flip-flip transitions, which is proportional to the difference in intensity of the electron spin resonance line shape and is called the differential solid effect. It always leads to a nuclear polarization well below the electron polarization.

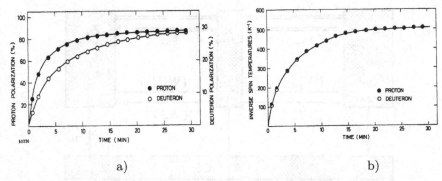

Fig. 4. a) Build-up of the proton and deuteron polarization (indicated on the left- and right-hand scale, resp.) in a partially deuterated sample. b) Build-up of the spin temperature of protons and deuterons.

the SS-reservoir. However, this method is unlikely to function well at low temperatures, since then all electron spins are in the lowest state, so there will be few double spin flips between the Zeeman levels of the electrons. Here came the excellent idea of Borghini:[4,5] he realized that the second method of a thermal contact is independent of the temperature, so it will be the dominant method at low temperature. So he extended the Provotorov rate equations[24] to include the nuclei and solved the three coupled differential equations for the temperatures of the SS- and Zeeman reservoirs of electrons and nuclei. The master equation was well explained by Borghini in his rejected thesis and I repeated the proof in the appendix of my thesis.[15] However, the formulae were written in the high temperature approximation, i.e. expanding the exponential function in the Boltzmann distribution. I extended the differential rate equations from Provotorov to low temperatures. The solutions could still be written analytically, but they were most easily solved numerically. Given that we obtained spin temperatures as low as a few μK, the precision had to be better than 10^{-10}, which I could nicely do on "my" HP2100.

3. Verifying the Mechanism of DNP in Polarized Targets

The first polarized targets consisted of frozen butanol beads doped with a free radical and reached a proton polarizations of about 40%.[16] A few years later propanediol doped with Cr-V complexes were used, in which a proton polarization close to 100% was obtained.[7,8] Such a high polarization would be impossible for the differential solid-state effect. So only the cooling via the SS-reservoir remained a possibility and we started a program to prove this. The predictions of Borghini's spin temperature model are clear: several nuclear species with different Larmor frequencies obtain a different polarization, but they have the same spin temperature, if the thermal contact is good enough, where good enough means that the leakage to the lattice is small in comparison with the heat transfer between the reservoirs in Fig. 3. This

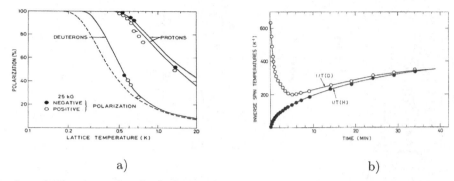

a)   b)

Fig. 5.   a) The observed and calculated polarization of protons and deuterons as function of the lattice temperature. b) Evolution of the spin temperature of protons and deuterons after destroying the proton polarization and switching on a microwave field to polarize again. The rapid equalization of the spin temperatures shows the increase of the thermal contact via the forbidden transitions (single arrows in Fig. 3). They finally reach the common temperature T_{SS} of the spin-spin interaction reservoir.

could be verified by observing the proton and deuteron polarization in a partially deuterated propanediol sample, as demonstrated in Fig. 4. During the polarization build-up the polarization becomes different (left side), but the spin temperature of the two nuclear species stays the same (right side). In the thermodynamical model of Fig 3 the final polarization depends on the leakage to the lattice, which is a strong function of temperature.[8] Fig. 5a shows the predicted and observed polarization of protons and deuterons in propanediol doped with Cr-V complexes as function of temperature.[13] Satisfactory agreement between theory and experiment is obtained. Here the heat capacities of the nuclear Zeeman reservoirs and temperature dependence of the spin-lattice relaxation times were carefully taken into account. The effect of the reduced heat capacity of the deuteron system is clearly seen by the difference between the dashed and solid line for the deuterons. At low temperatures the nuclear Zeeman reservoirs in Fig. 3 are rather isolated, at least without microwave irradiation establishing a thermal contact (single arrows in Fig. 3). This allows to destroy the polarization of one nuclear species by inducing spin transitions between the lower and upper nuclear Zeeman levels with a saturating RF field. If one e.g. destroys the proton polarization in a highly polarized sample, the deuteron polarization stays high. After switching on the microwave irradiation to polarize the sample again, this microwave irradiation establishes the thermal contact between the protons and deuterons, thus equalizing their spin temperatures much faster than expected from the polarization time by cooling of the SS reservoir. This is demonstrated in Fig. 5b, which clearly proves the dual role of the microwave field at low temperatures, the original idea of Michel.

Since the deuteron Larmor frequency is smaller than the width of the proton Zeeman levels, one has the same situation between deuterons and protons as for protons and electrons. Therefore, one should be able to polarize the deuterons by

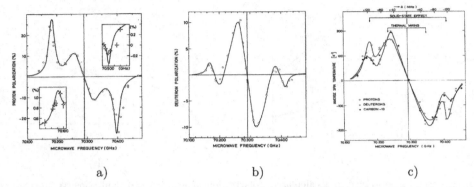

Fig. 6. a) Proton polarization as function of microwave frequency. b) Deuteron polarization as function of microwave frequency. c) Inverse spin temperature as function of microwave frequency.

off-center irradiation of a polarized proton spin system, thus cooling the SS-reservoir of the protons. Several schemes are possible: first polarize a sample of protons and deuterons by cooling the SS-reservoir of the electrons via off-center irradiation of the electron Zeeman levels with microwaves. Then destroy the deuteron polarization with a saturating RF field, followed by a cooling of the proton SS-reservoir with another RF field close to the proton Larmor frequency. Since the deuteron has S=1 the various levels can be populated such, that a pure tensor polarization can be obtained. Many experiments have been done and they all confirm the spin temperature theory in a quantitative way.[9, 11] Because of lack of space these beautiful experiments will not be described here.

4. Experiments Showing Different Mechanisms of DNP

As mentioned before, the solid-state mechanism of DNP is effective for narrow Zeeman energy levels of the unpaired electrons of the free radicals, while the mechanism via cooling of the spin-spin interaction reservoir is effective for electron Zeeman levels broader than the nuclear Zeeman splitting. The free radical BDPA has a width of the Zeeman levels, which is below the nuclear Zeeman levels of protons, but above the ones for deuterons and ^{13}C nuclei. Therefore one expects a combination of the two mechanisms, which should occur at different microwave frequencies. This was indeed the case as proven in a frozen sample of partially deuterated m-xylene (2,2-D6) doped with BDPA ($6 \cdot 10^{18}$ spins/cm^3). The measurements were performed in a magnetic field of 2.5T at a temperature of 0.75K.[12, 13, 15] The proton polarization is shown as function of microwave frequency in Fig. 6a. The inner peaks correspond to the polarization by the cooling of the SS-reservoir, while the outer peaks at frequencies $\nu_e \pm \nu_p$ correspond to the solid-state effect. The insets show the double solid-state effect at frequencies $\nu_e \pm 2\nu_p$ corresponding to a simultaneous spin flip of an electron and two protons. Deuterons have a Larmor frequency below the width of the electron Zeeman levels, so the polarization by the forbidden transitions of an electron and deuteron spin flip are not visible, since they are too close to the frequencies of the optimum cooling of the electron SS-reservoir. However, the double solid state effect of a simultaneous triple spin flip of an electron, proton and deuteron at frequencies $\nu_e \pm \nu_p \pm \nu_D$ leads to the four peaks in the deuteron polarization outside the main peak from the cooling of the SS-reservoir in Fig. 6b. The maximum polarization of protons and deuterons by cooling of the SS-reservoir (inner peaks in Figs. 6a and b) is about 10%. From a comparison with Fig. 4 it is clear that this does not correspond to an equal spin temperature, presumably because of the poor thermal contact between the proton Zeeman reservoir and the SS-reservoir. A good thermal contact requires the width of the electron Zeeman levels to be large with the nuclear Zeeman splitting, which is the case for the deuteron system, but not for the proton spin system. To check this hypothesis we prepared a sample of toluol-D8 with a larger concentration of BPA ($5 \cdot 10^{19}$ spins/cm^3), which increases the width of the electron Zeeman levels. In addition, we measured the polarization of the ^{13}C nuclei, which have a small Zeeman splitting as well and hence, should obtain the same spin temperature as the deuterons. This is indeed the case, as shown in Fig. 6c. The protons obtain indeed almost the same spin temperature with this sample with an order of magnitude higher concentration of BPDA leading to a broadening of the electron Zeeman levels.

5. Application of DNP in Cancer Therapy

Dynamic polarization has found an actively pursued application in cancer research: polarized ^{13}C nuclei in tracers of tumors yield a strongly enhanced signal in Magnetic Resonance Imaging (MRI), so smaller tumors can be discovered, see Ref.[25] for a recent development and references therein. The medical people call this hyperpolarization, but the polarization happens in setups similar to the ones used in polarized targets, see e.g..[26] The surprising discovery: after thawing the samples in a magnetic field, the polarization is largely maintained in the gas phase. Dissolving the gas into a liquid and injecting it into the body yields strongly enhanced NMR signals of the tumors. The polarization lasts only minutes, but this is enough for a picture in a modern magnetic resonance tomograph.

6. Summary

The mechanisms of dynamic polarization in polarized targets are by now well understood. According to Michel Borghini's idea this happens via a two-step process between the different heat reservoirs: i) cooling or heating of the SS-reservoir by off-center microwave irradiation; ii) establishing thermal contact between the SS-reservoir and the nuclear Zeeman reservoirs by the same microwave irradiation inducing triple spin-flips, namely a flip-flop transition of two electrons combined with a nuclear spin transition (either up or down until thermal equilibrium is reached). This mechanism was proven by many different experiments showing the thermal contact between the different reservoirs. By extending the usual high temperature approximation to low temperatures the spin temperature theory was proven to be valid to spin temperatures in the μK range, as was evident from the excellent agreement between theory and experiment. Nowadays the polarized targets, invented for high energy experiments, are used to enhance the polarization in biological tracers used to find tumors in Magnetic Resonance Imaging. The enhanced polarization provides a strongly enhanced signal, thus allowing to detect smaller tumors. The relatively high ^{13}C polarization of up to 60% is a clear manifestation, that Borghini's proposed mechanism of the dynamic polarization by cooling of the electron spin-spin interaction reservoir is at work. Michel certainly would have been delighted to see that his idea of dynamic polarization has found such important applications in fields never thought of before.

References

1. R. G. Milner, *PoS* **PSTP2013**, p. 3 (2013).
2. A. Krisch, *13th Workshop on High Energy Spin Physics (DSPIN-09), Dubna, Russia, C09-09-01* (2010).
3. R. de Beer, W. de Boer, C. van 't Hof and D. van Ormondt, *Acta Crystallogr.B Struct.Crystallogr.Cryst.Chem.* **29**, 1473 (1973).
4. M. Borghini, *Phys.Lett.* **26A**, p. 242 (1968).
5. M. Borghini, *Proc. 2nd International Conference on Polarized Targets, Berkeley, CA, USA* (Ed. G. Shapiro) **C710830**, 1 (1971).
6. M. Goldman and A. Landesman, *Phys.Rev.* **D132**, p. 610 (1963).
7. W. de Boer, *Nucl.Instrum.Meth.* **107**, 99 (1973).
8. W. de Boer and T. Niinikoski, *Nucl.Instrum.Meth.* **114**, 495 (1974).
9. W. de Boer, M. Borghini, K. Morimoto, T. Niinikoski and F. Udo, *Phys.Lett.* **B46**, 143 (1973).
10. M. Borghini, W. de Boer and K. Morimoto, *Phys.Lett.* **A48**, p. 244 (1974).
11. W. de Boer, *Phys.Rev.* **B12**, 828 (1975).
12. M. Borghini, W. de Boer and K. Morimoto, *Phys.Lett.* **A48**, p. 244 (1974).
13. W. de Boer, *J.Low.Temp.Phys.* **22**, p. 185 (1976).
14. A. Abragam and M. Goldman, *Rep. Prog. Phys.* **41**, p. 395 (1978).
15. W. de Boer, *PhD thesis, Tech. Univ. of Delft,* **CERN Yellow Report, CERN-74-11** (1974), http://cds.cern.ch/record/186203/files/CERN-74-11.pdf.
16. S. Mango, O. Runolfsson and M. Borghini, *Nucl.Instrum.Meth.* **72**, 45 (1969).
17. T. Niinikoski and F. Udo, *Nucl.Instrum.Meth.* **134**, p. 219 (1976).
18. W. de Boer, R. C. Fernow, A. Krisch, H. Miettinen, T. Mulera *et al.*, *Phys.Rev.Lett.* **34**, 558 (1975).
19. U. Amaldi, W. de Boer and H. Fürstenau, *Phys.Lett.* **B260**, 447 (1991).
20. W. de Boer, *Prog.Part.Nucl.Phys.* **33**, 201 (1994).
21. I. Solomon, *Proc. Magnetic and Electric Resonance and Relaxation, Amsterdam,* Ed. J. Schmidt , p. 25 (1963).
22. A. Redfield, *Phys.Rev.* **98**, p. 1787 (1955).
23. B. N. Provotorov, *JETP* **14**, p. 1126 (1961).
24. B. N. Provotorov, *JETP* **15**, p. 611 (1962).
25. T. Rodrigues, E. Serrao, B. W. C. Kennedy, D. Hu, K. M. I. and K. Brindle, *Nature Medicine* **20**, p. 9397 (2014).
26. T. Eichhorn, Y. Takado, N. Salameh et al., *Proc. of the Nat. Acad. of Science of the USA* **110(45)**, 18064 (2013).

The 21st International Symposium on Spin Physics (Spin2014)

Monday 20 October 2014 - Friday 24 October 2014

Peking University, Beijing, China.

Programme

Monday 20 October 2014

Opening and Plenary Session I (Chair: Haiyan Gao) - (08:15-10:00)

time	title	presenter
08:15	Welcome Remarks	XIE, Xincheng (Dean of School of Physics, Peking Univ.)
08:30	Spin Structure of the Nucleon	JI, Xiangdong (SJTU/UMD)
09:00	Parity-Violating Electron Scattering and the Weak Charge of the Proton	MACK, Dave (JLab)
09:30	Parity Violations in Hadronic Systems	SNOW, William (Indiana Univ.)

Symposium Photo and Coffee Break - (10:00-10:30)

Plenary Session II (Chair: Thomas Roser) - (10:30-12:00)

time	title	presenter
10:30	Beam Polarization at the ILC: Physics Case and Realisation	VAUTH, Annika (DESY)
11:00	Production of High Nuclear Spin Alignment of Radioactive Ion Beams	UENO, Hideki (RIKEN)
11:30	Physics with Polarized Targets in Storage Ring	TOPORKOV, Dmitriy (BINP and NSU)

Lunch Break - (12:00-13:30)

Parallel-I: S1 (Chair: Fan Wang) - (13:30-15:15)

time	title	presenter
13:30	Recent Developments in Nucleon Spin Decomposition	HATTA, Yoshitaka (Japan/Yukawa Institute)
14:00	Nucleon Tomography: Wigner Distributions	PASQUINI, Barbara (University of Pavia)
14:30	Extracting PDFs by Global Fit of Lattice QCD Calculations	MA, Yan-Qing (BNL)
14:55	Perturbative Matching of the Quasi-PDFs in Continuum Space and Lattice Space	YOSHIDA, Shinsuke (RIKEN)

Parallel-I: S5 (Chair: Xiaomei Li) - (13:30-15:15)

time	title	presenter
13:30	Transverse Spin Asymmetries in the CNI Region of Elastic Proton-Proton Scattering at $\sqrt{s}=200$ GeV	SVIRIDA, Dmitry (ITEP)
14:00	Measurement of the Analyzing Power in Proton-Proton Elastic Scattering at Small Angles	MACHARASHVILI, Giorgi (Joint Institute for Nuclear Research)
14:25	Study of the η Meson Production with Polarized Proton Beam	ZIELINSKI, Marcin (Jagiellonian University)
14:50	Initial Research of np Scattering with Polarized Deuterium Target at ANKE/COSY	GOU, Boxing (Institute of Modern Physics, Chinese Academy of Sciences)

Parallel-I: S8 (Chair: Jens Erler) - (13:30-15:15)

time	title	presenter

13:30	Recent Results and Progress on Leptonic and Storage Ring EDM Searches	KAWALL, David (University of Massachusetts Amherst)
14:00	Measurement of Muon g-2/EDM with Ultra-cold Muon at J-PARC	MIBE, Tsutomu (IPNS, KEK)
14:20	The New Muon g-2 Experiment at Fermilab	LI, Liang (Shanghai Jiao Tong University (CN))
14:40	High-precision Microwave Spectroscopy of Muonium for Determination of Muonic Magnetic Moment	TORII, Hiroyuki A. (University of Tokyo)

Parallel-I: S9 (Chair: Andreas Lehrach) - (13:30-15:15)

time	title	presenter
13:30	Polarimetry for Stored Polarized Hadron Beams	STEPHENSON, Edward (Indiana University)
13:55	High Precision Electron Beam Polarimetry	DUTTA, Dipangkar (Mississippi State University)
14:20	RHIC Proton Polarimetry Current Status and Future Plans	MAKDISI, Yousef (BNL)
14:38	Absolute and Relative Polarimeters for the SPACHARM Experiment.	SEMENOV, Pavel (IHEP)
14:56	The Deuteron Beam Polarimetry at Nuclotron-NICA	LADYGIN, Vladimir (Joint Institute for Nuclear Research (JINR))

Coffee Break - (15:15-15:30)

Parallel-II: S1 (Chair: Jianwei Qiu) - (15:30-17:30)

time	title	presenter
15:30	Hadron Structure from Drell-Yan Processes: An Overview	GROSSE-PERDEKAMP, Matthias (Univ. Illinois at Urbana-Champaign (US))
16:00	Overview of TMD Evolution	BOER, Daniel (University of Groningen)
16:30	Gluon TMDs and Quarkonium Production in Unpolarised and Polarised Proton-Proton Collisions	LANSBERG, Jean-Philippe (IPN Orsay, Paris Sud U. / IN2P3-CNRS)
16:50	Some New Opportunities for Spin Physics at Small x	ZHOU, Jian (Regensburg University)

Parallel-II: S5 (Chair: Yajun Mao) - (15:30-17:30)

time	title	presenter
15:30	Spin Physics with PHENIX Experiment's MPC-EX Calorimeter Upgrade.	JIANG, Xiaodong (Los Alamos National Laboratory)
16:00	Azimuthal Asymmetries of Drell-Yan Process in pA Collisions	GAO, Jian-Hua (Shandong University at Weihai)
16:25	High Twist Effects in $e^+ e^-$ Annihilation at High Energies	WEI, Shu-yi (School of Physics, Shandong University)

Parallel-II: S8 (Chair: Krishna Kumar) - (15:30-17:30)

time	title	presenter

15:30	Parity Violation in Deep Inelastic Scattering with the SoLID Spectrometer at JLab	SOUDER, Paul (Syracuse University)
16:00	Searches for Physics beyond the Standard Model at the LHC	ELLINGHAUS, Frank (Johannes-Gutenberg-Universitaet Mainz (DE))
16:30	Towards a Precision Measurement of the Muon Pair Asymmetry in e+e- Annihilation at Belle and Belle II	FERBER, Torben (DESY)
16:50	Test of Time-Reversal Invariance at COSY (TRIC)	EVERSHEIM, Dieter (Helmholtz Institut für Strahlen- und Kernphysik, University Bonn, Germany)
17:10	Test of Time-reversal Symmetry in the Proton - Deuteron Scattering	UZIKOV, Yuriy (Joint Institue for Nuclear Researches)

Parallel-II: S9 (Chair: Mei Bai) - (15:30-17:30)

time	title	presenter
15:30	Effect of Overlapping Intrinsic Spin Resonances on e-lens lattices from FY13 Polarized Proton Run	RANJBAR, Vahid (BNL)
15:55	High energy Polarized Electrons	SHATUNOV, Yury (Budker institute of nuclear physics)
16:20	Polarization Preservation and Control in a Figure-8 Ring	MOROZOV, Vasiliy (Jefferson Lab, Newport News, VA, USA)
16:45	Beam Polarization Aspects of eRHIC	PTITSYN, Vadim (Brookhaven National Laboratory)
17:05	Study of the Polarization Deterioration During Physics Stores in RHIC Polarized Proton Runs	DUAN, Zhe (institute of high energy physics, Chinese Academy of Sciences)

Parallel-II: S12 (Chair: Xincheng Xie) - (15:30-17:30)

time	title	presenter
15:30	Quantum Computation and Quantum Metrology Based on Solid Spin System	DU, Jiangfeng (USTC)
16:00	Quantum Information Processing Based on NV Centers	PAN, Xin-Yu (IoP, CAS)
16:30	Quantum Computing and Entanglement Purification on Electron Spins	DENG, Fu-Guo (Beijing Normal Univ.)
17:00	Quantum Simulation with Nuclear Spins	LONG, Gui-Lu (Tsinghua Univ.)

Tuesday 21 October 2014

▌Plenary Session III (Chair: Oleg Teryaev) - (08:00-10:00)

time	title	presenter
08:00	Lattice QCD for Spin Physics	LIU, Keh-Fei (Univ. of Kentucky)
08:30	Nucleon Electromagnetic Form Factors	CATES, Gordon (Univ. of Virginia)
09:00	Latest Results from the COMPASS Experiment	STOLARSKI, Marcin (LIP)
09:30	Results from RHIC Spin Program	RENEE, Fatemi (Univ. of Kentucky)

Coffee Break - (10:00-10:15)

▌Plenary Session IV (Chair: Franco Bradamante) - (10:15-12:00)

time	title	presenter
10:15	Highlights from HERMES	ROSTOMYAN, Armine (DESY)
10:45	Summary of PSTP2013	POELKER, Matthew (JLab)
11:00	Summary of D-SPIN2013	EFREMOV, Anatoly V. (JINR)
11:15	Summary of SPIN-Praha-2013	FINGER, Michael (Prague)
11:30	Medical Imaging with Highly Spin Polarized Molecules	WARREN, Warren (Duke Univ.)

Lunch Break - (12:00-13:30)

▌Parallel-III: S2 (Chair: Jian-Ping Chen) - (13:30-15:15)

time	title	presenter
13:30	Theoretical Status of Helicity Parton Densities	NOCERA, Emanuele Roberto (Università degli Studi di Milano & INFN Milano,Italy)
14:00	New Results on the Longitudinal Spin Structure of the Nucleon from CLAS at Jefferson Lab	BOSTED, Peter (College of William and Mary)
14:30	COMPASS Results on g1 and NLO QCD Fits	KUNNE, Fabienne (CEA/IRFU,Centre d'etude de Saclay Gif-sur-Yvette (FR))

▌Parallel-III: S4 (Chair: Nicole D'Hose) - (13:30-15:15)

time	title	presenter
13:30	Nucleon Form Factors	BRISCOE, Bill (George Washington Univ.)
14:00	Overview of the Proton Radius Problem	CARLSON, Carl (William and Mary)
14:30	New Precision Measurement for Proton Zemach Radius with Laser Spectroscopy of Muonic Hydrogen	MA, Yue (RIKEN)

Parallel-III: S5 (Chair: Nikolai Kochelev) - (13:30-15:15)

time	title	presenter
13:30	Role of Spin in NN → NNπ	BARU, Vadim (Ruhr University Bochum)
14:00	Spin Physics at NICA	NAGAYTSEV, Alexander (JINR Dubna on behalf SPD team)

Parallel-III: S8 (Chair: Wei-Tou Ni) - (13:30-15:15)

time	title	presenter
13:30	Spin-gravity Interactions and Equivalence Principle	TERYAEV, Oleg (JINR)
14:18	New Limit on Lorentz-Invariance- and CPT-Violating Neutron Spin Interactions Using a Free-Spin-Precession ^{3}He-^{129}Xe Comagnetometer	ALLMENDINGER, Fabian (Physikalisches Institut, Uni Heidelberg, Germany)
14:37	Limits for Spin-Dependent Short-Range Interaction of Axion-like Particles	TULLNEY, Kathlynne (University Mainz, Germany)
14:56	New Spin and Velocity Dependent Force Searching by Using Polarized Helium Gas	FU, Changbo (Shanghai Jiaotong University)

Parallel-III: S12 (Chair: Gui Lu Long) - (13:30-15:15)

time	title	presenter
13:30	Planar Quantum Squeezing and Optimized Interferometric Phase Measurement	HE, Qiongyi (Peking Univ.)
14:00	Quantum Information Processing Based on Quantum-dots in Optical Double-sided Microcavities	WANG, Tie-jun (Beijing U Post and Communications)
14:25	Demonstrating a Quantum Algorithm Using Nuclear Magnetic Resonance	LIU, Yang (North China Electric Power University)

Coffee Break - (15:15-15:30)

Parallel-IV: S2 (Chair: Renee Fatemi) - (15:30-17:30)

time	title	presenter
15:30	Proton Quark Helicity Structure via W-boson Production in pp Collision at \sqrt(s) = 500 GeV @ PHENIX	GIORDANO, Francesca (UIUC)
16:00	Proton Spin Content in Lattice QCD from a Large Momentum Effective Field Theory	ZHAO, Yong (University of Maryland)
16:15	Measurement of Longitudinal Spin Asymmetries for Weak Boson Production in Polarized Proton-Proton Collisions at STAR	ZHANG, Jinlong (Shandong University/Brookhaven National Laboratory)
16:30	Symmetry Breaking and Determination of Strange Quark Distribution of the Nucleon	CAO, Fu-Guang (Massey University)
16:45	Accessing Polarized Sea Flavor Asymmetry through Semi-Inclusive DIS at JLab-12GeV and the Future EIC	JIANG, Xiaodong (Los Alamos National Laboratory)

Parallel-IV: S4 (Chair: Barbara Pasquini) - (15:30-17:30)

time	title	presenter
15:30	Nucleon Form Factors: An Incisive Window into Quark-Gluon Dynamics	CLOET, Ian (Argonne National Laboratory)

16:00	Theory and Phenomenology of GPDs	KUMERICKI, Kresimir (University of Zagreb)
16:30	Progress in Double Distributions and Generalized Parton Distributions Modeling.	MEZRAG, Cedric (IRFU/SPhN)
16:50	Generalized Parton Distributions for the Nucleon in the Soft-wall Model of AdS/QCD	SHARMA, Neetika (Indian institute of Science Education and Research Mohali)

Parallel-IV: S6 (Chair: Reinhard Beck) - (15:30-17:30)

time	title	presenter
15:30	Double Polarisation Experiments in Meson Photoproduction and the Impact on the Nucleon Excitation Spectrum	HARTMANN, Jan (HISKP, Univerity of Bonn)
16:00	Double Polarisation Experiments in Meson Photoproduction at JLab	PASYUK, Eugene (Jefferson Lab)
16:30	Spin Density Matrix Elements in Exclusive Production of Omega Mesons at HERMES	MARUKYAN, Hrachya (DESY)

Parallel-IV: S9 (Chair: Edward Stephenson) - (15:30-17:30)

time	title	presenter
15:30	Storage Ring Based EDM Search	LEHRACH, Andreas (Forschungszentrum Juelich)
15:55	Systematic Calculation of Spin Resonance Strengths to High Order	BARBER, Desmond (Deutsches Elektronen Synchrotron (DESY), Germany.)
16:15	Studies of Systematic Limitations in the EDM Searches at Storage Rings	SALEEV, Artem (Forschungszentrum Juelich)
16:30	A Novel RF-ExB Spin Manipulator at COSY (Jülich, Germany)	MEY, Sebastian (Forschungszentrum Juelich)
16:45	Machine Development for Spin	LENISA, Paolo (University of Ferrara and INFN)

Parallel-IV: S12 (Chair: Warren Warren) - (15:30-17:30)

time	title	presenter
15:30	Dipolar and Quadrupolar Signatures of Topological Band Structures	BOUCHARD, Louis (UCLA)
16:00	Polarized Fusion: Can Polarization Help to Increase the Energy Output of Fusion Reactors?	ENGELS, Ralf (FZ Jülich)
16:25	Polarized Fusion and its Implications: The Potential for Direct In situ Measurements of Fuel Polarization Survival in a Tokamak Plasma	SANDORFI, Andrew (Jefferson Lab)
16:50	Generation of States with Long-lived Molecular Polarization and Catalytic Generation In-Magnet of Molecular Spin Hyperpolarization	WARREN, Warren (Duke Univ.)
17:10	Recent Trends in Laser-polarized Gases for Medical Imaging and Nuclear Targets	CATES, Gordon (Univ. of Virginia)

Wednesday 22 October 2014

Parallel-V: S2 (Bo-Wen Xiao) - (08:15-10:15)

time	title	presenter
08:15	Constraints on ΔG from COMPASS Data	KUREK, Krzysztof (National Centre for Nuclear Research (PL))
08:45	Gluon Polarization in Longitudinally Polarized pp Collisions at STAR	CHANG, Zilong (Texas A&M University)
09:15	Impact of PHENIX Measurements of $A_{LL}^{\pi^0}$ on Determination of the Gluon Spin in the Proton	MANION, Andrew (Lawrence Berkeley National Laboratory)
09:30	Recent Results of $A_{LL}^{\pi^0}$ Measurements at $\sqrt{s} = 510$ GeV at Mid-rapidity by PHENIX Experiment and Resulting Constraint on the Gluon Spin Contribution to the Proton Spin	YOON, Inseok (Seoul National University)
09:45	J/ψ Longitudinal Double Spin Asymmetry Measurement at Forward Rapidity in $p+p$ Collisions at $\sqrt{s}=510$ GeV	YU, Haiwang (PKU, NMSU, LANL)
10:00	Double Helicity Asymmetries of Forward Neutral Pions from $\sqrt{s}=510$ GeV pp Collisions at STAR	DILKS, Christopher (Pennsylvania State University)

Parallel-V: S3 (Chair: Xiang-Song Chen) - (08:15-10:15)

time	title	presenter
08:15	Transverse Momentum and Spin Dependent Distribution Functions	MULDERS, Piet J. (VU/Nikhef)
08:45	Recent Key Measurements for Accessing the Transverse Spin and Momentum Structure of the Nucleon	MARTIN, Anna (Trieste University and INFN (IT))
09:15	Transverse Spin Azimuthal Asymmetries in SIDIS at COMPASS	PARSAMYAN, Bakur (University of Turin and INFN (IT))
09:30	Transversity Experiment (E06-010) at Jlab	ZHAO, Yuxiang (USTC)
09:45	Transverse Single-spin Asymmetries of Pion Production in Semi-inclusive DIS at Subleading Twist	MAO, Weijuan (Southeast University)
10:00	Momentum Structure of the Nucleon and Hadron Production in Unpolarised SIDIS at COMPASS	MAKKE, Nour (Universita e INFN (IT))

Parallel-V: S6 (Chair: Xiaodong Jiang) - (08:15-10:15)

time	title	presenter
08:15	Spin Physics with Photons - Technical Highlights and Spin-offs	THOMAS, Andreas (University Mainz)
08:45	Spin Degrees of Freedom in Compton Scattering	MISKIMEN, Rory (University of Massachusetts)
09:15	Spin Observables in Pion Photoproduction and Nucleon Compton Scattering from the Chiral Lagrangian and Dispersion Relations.	GASPARYAN, Ashot (Ruhr University of Bochum)
09:35	Transverse and Longitudinal Lambda Polarization in Lepton Scattering by Unpolarized Nucleons at HERMES	KARYAN, Gevorg (DESY)

Parallel-V: S9 (Chair: Vasiliy Morozov) - (08:15-10:15)

time	title	presenter

08:15	NICA Facility In Polarized Proton and Deuteron Mode	KOVALENKO, Alexander (Joint Institute for Nuclear Research)
08:40	Multiple Horizontal Tune Jumps To Overcome Horizontal Depolarizing Resonances	HUANG, Haixin (Brookhaven National Lab)
08:58	A New Formalism for Classifying Spin Motions Using Tools Distilled from the Theory of Principal Bundles	BARBER, Desmond (Deutsches Elektronen Synchrotron (DESY), Germany)
09:16	Calculation of Spin Resonances with Strong Betatron Coupling	PTITSYN, Vadim (Brookhaven National Laboratory)
09:34	Estimation of Systematic Errors for Deuteron Electric Dipole Moment (EDM) Search at a Storage Ring	CHEKMENEV, Stanislav (Rheinisch-Westfälische Technische Hochschule (RWTH))
09:52	Electron Polarization In The MEIC Collider Ring At JLab	LIN, Fanglei (Thomas Jefferson National Accelerator Facility)

Coffee Break - (10:15-10:30)

Parallel-VI: S2 (Chair: Peter Bosted) - (10:30-12:00)

time	title	presenter
10:30	Study of g2 Spin Structure Function at Jefferson Lab	CHOI, Seonho (Seoul National University)
11:00	A Measurement of g_2^p at Low Q^2	CUMMINGS, Melissa (The College of William and Mary)
11:15	Quark to the Λ and $\bar \Lambda$ Spin Transfers in the Current Fragmentation Region	DU, Xiaozhen (Peking University)
11:30	COMPASS Results on Hadron Multiplicities and Fragmentation Functions	KUNNE, Fabienne (CEA/IRFU,Centre d'etude de Saclay Gif-sur-Yvette (FR))
11:45	Multiplicities of Charged Pions and Kaons in Deep-inelastic Scattering by Protons and Deuterons at HERMES and the Strange-quark Distribution in the Nucleon	KARYAN, Gevorg (DESY)

Parallel-VI: S3 (Chair: Matthias Grosse-Perdekamp) - (10:30-12:00)

time	title	presenter
10:30	Fragmentation Function Measurements at Belle	GIORDANO, Francesca (UIUC)
10:45	New COMPASS Results on Transverse Spin Asymmetries in Hadron Pair Production in DIS	SBRIZZAI, Giulio (Trieste Univ. and INFN)
11:00	QCD Evolution Effects for the Collins Azimuthal Asymmetries in e+e- and SIDIS	SUN, Peng (LBNL)
11:15	Measurement of Double Collins Asymmetries at BESIII	GUAN, Yinghui (IHEP)
11:30	Phenomenological Extraction of Transversity from COMPASS SIDIS and Belle e+e- Data	BRADAMANTE, Franco (Universita e INFN (IT))

Parallel-VI: S6 (Chair: Reinhard Beck) - (10:30-12:00)

time	title	presenter

| 10:30 | Complete Experiments in Meson Photoproduction | WUNDERLICH, Yannick (University of Bonn) |
| 11:00 | Unraveling Excitations of the Nucleon – Meson Photo-production from Polarized Neutrons in HD at CLAS | SANDORFI, Andrew (Jefferson Lab) |

Parallel-VI: S7 (Chair: Zhigang Xiao) - (10:30-12:00)

time	title	presenter
10:30	Novel Spin Modes in Exotic Nucleus	MENG, Jie (Peking University)
11:00	N3LO Chiral Predictions for Spin Observables in Nucleon-Deuteron Elastic Scattering and the Deuteron Breakup at Low Energies.	SKIBINSKI, Roman (Jagiellonian University)
11:20	Study of Three Nucleon Force Effects via Few-Nucleon Scattering	WADA, Yasunori (Tohoku university)
11:40	Polarization Effects in Deuteron-Induced Reactions	OU, Li (Guangxi Normal University)

Parallel-VI: S10 (Chair: Matt Poelker) - (10:30-12:00)

time	title	presenter
10:30	Review of Polarized Ion Sources	ZELENSKI, Anatoli (BNL)
11:00	Design of Transversal Phase Space Meter for Atomic Hydrogen Beam Source	BELOV, Alexander (INR)
11:20	Polarized He3 Ion Source for RHIC and eRHIC	MAXWELL, James (MIT)
11:40	Status of the New Source of Polarized Ions for the JINR Accelerator Complex	FIMUSHKIN, Victor (Joint Institute for Nuclear Research)

Parallel-VI: S11 (Chair: Jianwei Qiu) - (10:30-12:00)

time	title	presenter
10:30	Systematic Study of Spin Effects at SPASCHARM Experiment	MOCHALOV, Vasily (IHEP, Protvino)
10:50	New p+p and p+A Physics Opportunities with the Forward sPHENIX Upgrade at the Relativisitc Heavy Ion Collider	LIU, Ming Xiong (Los Alamos National Laboratory)
11:10	MPD and BM@N Detectors at NICA. Prospects for the Polarization Effects Measurements.	PESHEKHONOV, Dmitry (Joint Institute for Nuclear Research)

Lunch Break - (12:00-13:30)

Excursion - (13:30-17:30)

Public Lecture - (19:00-20:00)

time	title	presenter
19:00	Quantum Anomalous Hall Effect and Information Technology	XUE, Qi-Kun (Tsinghua Univ.)

Thursday 23 October 2014

Plenary Session V (Chair: Anna Martin) - (08:15-10:15)

time	title	presenter
08:15	Three-dimensional Imaging of the Nucleon: TMD (Theory/Phenomenology)	LIANG, Zuo-Tang (Shandong Univ.)
08:45	Spin Physics with 12-GeV CEBAF	ENT, Rolf (JLab)
09:15	Future Spin Physics at RHIC	BOYLE, Kieran (BNL)
09:45	EPJ A Lecture: Spin Physics at an Electron-Ion Collider	MEZIANI, Zein-Eddine (Temple Univ.)

Coffee Break - (10:15-10:30)

Plenary Session VI (Chair: Hans Stroeher) - (10:30-12:00)

time	title	presenter
10:30	Spin Physics at J-PARC	KUMANO, Shunzo (KEK)
11:00	Spin Physics at COSY - Recent Results and Future Plans	KACHARAVA, Andro (Forschungszentrum Juelich)
11:30	Latest Results on Few-body Physics	AHMED, Mohammad (NCCU and TUNL)

Lunch Break - (12:00-14:00)

Memorial Session in Honor of M. Borghini (Chair: Alan Krisch) - (14:00-16:00)

time	title	presenter
14:00	Introduction	KRISCH, Alan D. (University of Michigan)
14:15	The Dawn of High Energy Spin Physics	MASAIKE, Akira (Kyoto Univ.)
14:55	Today's Polarized Targets in Borghini's Footsteps	MEYER, Werner Peter (Ruhr-Universitaet Bochum (DE))
15:25	Borghini's Contributions to Today's Polarized Targets	CRABB, Donald G. (University of Virginia)
15:50	Borghini as a mentor at CERN	

Poster Session (Chair: Xiaomei Li) - (14:00-16:00)

time	title	presenter
15:00	Coffee Hour	

Banquet - (18:30-21:00)

Friday 24 October 2014

Parallel-VII: S3 (Chair: Han-Xin He) - (08:15-10:15)

time	title	presenter
08:15	Key Future Measurements of TMDs at JLab and Other Facilities	ALLADA, Kalyan (Massachusetts Institute of Technology)
08:45	Transverse Target Single-spin Asymmetry in Inclusive Electroproduction of Charged Pions and Kaons	SCHNELL, Gunar (DESY)
09:00	Azimuthal Asymmetries for eA/eN Semi-inclusive DIS and Its Nuclear Dependence	SONG, Yu-kun (Jinan Univ.)
09:15	Transverse Single Spin Asymmetries of Forward π^{0} and Jet-like Events in $\sqrt{s}$ = 500 GeV Polarized Proton Collisions at STAR	PAN, Yuxi (University of California, Los Angeles)
09:30	Measuring Transversity with Di-hadron Correlations in p$\{^{\uparrow}\}$+p Collisions at $\sqrt{s}$ = 500 GeV	SKOBY, Michael (Indiana University)
09:45	New COMPASS Work on the interplay among h+, h-, and 2h transverse spin asymmetries in SIDIS	BRADAMANTE, Franco (trieste university and infn)
10:00	Azimuthal Single-Spin Asymmetries of Charged Pions in Jets in $\sqrt{s}=200$ GeV $p^{\uparrow}p$ Collisions at STAR	ADKINS, J. Kevin (University of Kentucky)

Parallel-VII: S4 (Chair: Nicole D'Hose) - (08:15-10:15)

time	title	presenter
08:15	An Experimental Overview on DVCS Measurements (Past, Present and Future)	FISCHER, Horst (Albert-Ludwigs-Universitaet Freiburg (DE))
08:45	DVCS at HERMES	MARUKYAN, Hrachya (DESY)
09:05	The DVCS Physics Program at COMPASS	FERRERO, Andrea (Commissariat a l'Energie Atomique et aux Energies Alternatives)
09:25	Deeply Virtual Meson Production at Jefferson Lab	KUBAROVSKY, Valery (Jefferson Lab)
09:45	Hard Exclusive Meson Production at COMPASS	TER WOLBEEK, Johannes (Albert-Ludwigs-Universitaet Freiburg (DE))

Parallel-VII: S7 (Chair: Kimiko Sekiguchi) - (08:15-10:15)

time	title	presenter
08:15	Studies of Unstable Nuclei with Spin-Polarized Proton Target	SAKAGUCHI, Satoshi (Kyushu University)
08:45	Polarized Proton Elastic Scattering and Nucleon Density Distributions	ZENIHIRO, Juzo (RIKEN)
09:05	Relativistic Studies of Spin-isospin Response in Nuclei	LIANG, Haozhao (RIKEN Nishina Center)
09:25	Recent Progress in Spin-isospin Excitations in Nuclei	HIROYUKI, Sagawa (RIKEN/University of Aizu)
09:45	Spin Polarization of Radioisotope Atoms with Optical Pumping in Superfluid Helium for the Measurement of Nuclear Spins and Electromagnetic Moments	FURUKAWA, Takeshi (Department of Physics, Tokyo Metropolitan University)

Parallel-VII: S10 (Chair: Don Crabb) - (08:15-10:15)

time	title	presenter
08:15	Review of Solid Polarized Targets	KEITH, C. (JLab)
08:45	High Deuteron Polarization in Polymer Target Materials	WANG, Li (Donghua University)
09:00	Tensor Polarization Optimization and Measurement for Solid Spin 1 Targets	KELLER, Dustin (University of Virginia)
09:20	A New Polarized Solid Proton Target for Fermilab E1039 Drell-Yan Experiment	LIU, Ming (Los Alamos National Laboratory)
09:35	Preparations for Electron Beam Experiments with Transversely Polarized Solid HD	WEI, Xiangdong (Jefferson Lab)
09:55	The H and D Polarized Target for Spin-filtering Measurements at COSY.	CIULLO, Giuseppe (INFN and Dipartimento di Fisica e Scienze della Terra - 44121 Ferrara Italy)

Parallel-VII: S11 (Chair: Rolf Ent) - (08:15-10:15)

time	title	presenter
08:15	EIC in the US	BAI, Mei (BNL)
08:45	EIC/HIAF in China	CHEN, Xurong (IMP)
09:15	ePHENIX: An Electron Ion Collider (EIC) Detector Built Around the BaBar Magnet	BAZILEVSKY, Alexander (Brookhaven National Laboratory)
09:35	TMD Studies with A Fixed-Target ExpeRiment Using the LHC Beams (AFTER@LHC)	MASSACRIER, Laure Marie (LAL/IPNO)
09:55	TMD Studies and More with SoLID at JLab	CHEN, Jian-Ping (JLab)

Parallel-VII: S12 (Chair: Xi Chen) - (08:15-10:15)

time	title	presenter
08:15	Spin Superconductor and Electric Dipole Superconductor	XIE, Xincheng (Peking Univ.)
08:50	Aharonov-Casher Effect in Bi2Se3 Square-Ring Interferometers	LU, Li (Institute of Physics, Chinese Academy of Sciences)
09:25	Real-space Imaging of Dirac-Landau Orbits in Topological Surface State	FU, Yingshuang (Huazhong Univ. of Science & Technology)
09:45	Spin-wave Nanograting Coupler	YU, Haiming (Beihang Univ.)

Coffee Break - (10:15-10:30)

Parallel-VIII: S3 (Chair: Ming Liu) - (10:30-12:00)

time	title	presenter
10:30	Measurement of Boer-Mulders Function via Drell-Yan Process by SeaQuest Experiment at Fermilab	NAKANO, Kenichi (Tokyo Tech)
10:45	Working With Wilson Lines	VAN DER VEKEN, Frederik (University of Antwerp)
11:00	Transverse Single-spin Asymmetry of Heavy-flavor Production in $p+p$ Collisions at $\sqrt{s}$ = 200 GeV	WEI, Feng (New Mexico State University)

11:15	Gluon Contribution to the Sivers Effect COMPASS Results on Deuteron Target.	KUREK, Krzysztof (National Centre for Nuclear Research (PL))
11:30	Transverse Single Spin Asymmetry of π^0 and η Mesons at RHIC/PHENIX	XIAORONG, Wang (New Mexico State University and RBRC)
11:45	Single-spin Asymmetries in SIDIS on the Longitudinally Polarized Nucleon Targets	LU, Zhun (Southeast University)

Parallel-VIII: S4 (Chair: Chuan Liu) - (10:30-12:00)

time	title	presenter
10:30	Nucleon Structure from Lattice QCD	ZANOTTI, James (University of Adelaide)
11:00	Parton Distribution Functions and Matching	ZHANG, Jianhui (SJTU)
11:20	Quark Angular Momentum in a Spectator Model	LIU, Tianbo (Peking University)
11:40	Quark Wigner Distributions and Orbital Angular Momentum in Light-front Dressed Quark Model	NAIR, Sreeraj (Indian Institute of Technology, Bombay)

Parallel-VIII: S7 (Chair: Seonho Choi) - (10:30-12:00)

time	title	presenter
10:30	Polarization Observables in Few-Nucleon Scattering	KISTRYN, Stanislaw (Jagiellonian University Krakow)
11:00	Development of Neutron Polarization Measurement System for Studying NN Interaction in Nuclear Medium	YASUDA, Jumpei (Department of Physics, Kyushu University)
11:20	Spin-orbit Splitting of Oxygen Isotopes Studied via (pol p, 2p) Reaction	KAWASE, Shoichiro (the University of Tokyo)
11:40	The Spin Studies in Few Body Systems at Nuclotron	LADYGIN, Vladimir (LHEP-JINR)

Parallel-VIII: S10 (Chair: Dipangkar Dutta) - (10:30-12:00)

time	title	presenter
10:30	High Current Electron Guns for eRHIC	BEN-ZVI, Ilan (Brookhaven National Laboratory)
11:00	DC-SRF Photo-cathode Gun at Peking University	LIU, Kexin (Peking University)
11:25	Overview of Polarized He3 Gas Targets	CHEN, Jian-ping (Jefferson Lab)

Parallel-VIII: S11 (Chair: Alexander Nagaytsev) - (10:30-12:00)

time	title	presenter
10:30	Opportunities with Polarized Hadron Beams	LORENZON, Wolfgang (Michigan)
11:00	Polarized Drell-Yan at COMPASS-II: Transverse Spin Physics Program	PARSAMYAN, Bakur (University of Turin and INFN (IT))
11:20	Cross Section and Asymmetry Measurement of Very Forward Neutral Particle Production at RHIC	GOTO, Yuji (RIKEN)

| 11:40 | Nucleon Partonic Spin Structure to be explored by the Unpolarized Drell-Yan Program in the COMPASS-II Experiment at CERN | CHANG, Wen-Chen (Academia Sinica (TW)) |

Parallel-VIII: S12 (Chair: Li Lu) - (10:30-12:00)

time	title	presenter
10:30	Quantum Anomalous Hall Effect in Magnetically Doped Topological Insulator	HE, Ke (Tsinghua Univ.)
11:00	U(1) Symmetry Protected Spin Quantum Hall Effect in S = 1 Gutzwiller Wavefunction	LIU, Zhengxin (Tsinghua Univ.)
11:20	Observation of the Surface States in Topological Kondo Insulator SmB6 and YbB6	ZHANG, Tong (Tsinghua Univ.)
11:40	Dynamical Generation of Fermion Mass in Weyl Semimetals	WANG, Zhong (Tsinghua Univ.)

Lunch Break - (12:00-13:30)

Plenary Session VII (Chair: Erhard Steffens) - (13:30-15:45)

time	title	presenter
13:30	Searches for EDMs	FILIPPONE, Brad (Caltech)
14:00	Low-Energy Tests of Standard Model	MAAS, Frank (Mainz)
14:30	Search for the Role of Spin and Polarization in Gravity	NI, Wei-Tou (National Tsinghua Univ.)
15:00	Symposium Summary	MILNER, Richard (MIT)
15:30	Symposium Closing	

The 21st International Symposium on Spin Physics (Spin2014)

Monday 20 October 2014 - Friday 24 October 2014

Peking University, Beijing, China.

List of Participants

Name	Email	Institution	City	Country
Dr. ABRAMOV, Victor	victor.abramov@ihep.ru	Institute for High Energy Physics	Protvino	Russian Federation
Mr. ADKINS, James	kevin.adkins@uky.edu	University of Kentucky	Lexington	United States
Dr. AHMED, Mohammad	ahmed@tunl.duke.edu	TUNL and North Carolina Central University	Durham, NC	United States
Dr. ALLADA, Kalyan	kalyan@jlab.org	Massachusetts Institute of Technology	Newport News	United States
ALLMENDINGER, Fabian	allmendinger@physi.uni-heidelberg.de	Physikalisches Institut, Uni Heidelberg, Germany	Heidelberg	Germany
Dr. BAI, Mei	mbai@bnl.gov	Brookhaven National Laboratory	Upton	United States
Prof. BARBER, Desmond	mpybar@mail.desy.de	DESY	Hamburg	Germany
Dr. BARU, Vadim	vbaruru@gmail.com	Ruhr University Bochum	Bochum	Germany
Dr. BAZILEVSKY, Alexander	shura@bnl.gov	Brookhaven National Laboratory	Upton	United States
Prof. BECK, Reinhard	reinhard.beck@cern.ch	Universitaet Bonn (DE)	Bonn	Germany
Dr. BELOV, Alexander	alexandre.belov@cern.ch	INR Russian Academy of Sciences (RU)	Moscow	Russian Federation
Prof. BEN-ZVI, Ilan	benzvi@bnl.gov	Brookhaven National Laboratory	Upton	United States
Prof. BOER, Daniel	d.boer@rug.nl	University of Groningen	Groningen	Netherlands
Prof. BOSTED, Peter	bosted@jlab.org	College of William and Mary	Williamsburg	United States
Prof. BOUCHARD, Louis	louis.bouchard@gmail.com	UCLA	Los Angeles	United States
Dr. BOYLE, Kieran	kboyle@bnl.gov	RIKEN BNL Research Center	Upton	United States
Dr. BOYLE, Kieran	kpboyle@gmail.com	RIKEN BNL Research Center	Upton, NY	United States
Prof. BRADAMANTE, Franco	franco.bradamante@cern.ch	Universita e INFN (IT)	Trieste	Italy
Dr. BRESSAN, Andrea	andrea.bressan@cern.ch	Universita e INFN (IT)	Trieste	Italy
Prof. BRISCOE, William John	briscoe@gwu.edu	The George Washington University	Washington	United States
Prof. BURINSKII, Alexander	burinskii@mail.ru	NSI Russian Academy of Sciences	Moscow	Russian Federation

Name	Email	Institution	City	Country
Dr. CAO, Fu-Guang	f.g.cao@massey.ac.nz	Massey University	Palmerston North	New Zealand
Prof. CARLSON, Carl E.	carlson@physics.wm.edu	College of William and Mary	Williamsburg, VA 23187	United States
Prof. CATES, Gordon	cates@virginia.edu	University of Virginia	Charlottesville, Virginia	United States
Prof. CHANG, Wen-Chen	changwc@phys.sinica.edu.tw	Institute of Physics, Academia Sinica	Taipei	Taiwan
Mr. CHANG, Zilong	changzl@physics.tamu.edu	Texas A&M University	College Station, TX	United States
Mr. CHEKMENEV, Stanislav	s.chekmenev@fz-juelich.de	Rheinisch-Westfälische Technische Hochschule (RWTH) III.Physikalisches Institut B Physikzentrum	Aachen	Germany
Prof. CHEN, Jian-ping	jpchen@jlab.org	Jefferson Lab	Newport News, Virginia	United States
CHEN, Kaibao	chenkaibao@gmail.com	Shandong University	Jinan	China
CHEN, Xiang-Song	cxs@hust.edu.cn	Huazhong University of Science and Technology	Wuhan	China
Prof. CHEN, Xurong	xchenimp@gmail.com	Institute of Modern Physics, CAS	Lanzhou	China
Mr. CHEN, boping	372064995@qq.com	Huazhong University of Science & Technology	Wuhan	China
Dr. CHERNITSKII, Alexander	aachernitskii@mail.ru	Saint-Petersburg State University of Economics	Saint-Petersburg	Russian Federation
Prof. CHOI, Seonho	leojustin@gmail.com	Seoul National University	Seoul, 151-747	Korea, Republic of
Dr. CIUFFOLI, Emilio	ciuffoli@ihep.ac.cn	IHEP, CAS	Beijing	China
Dr. CIULLO, Giuseppe	ciullo@fe.infn.it	INFN-Ferrara and Dipartimento di Fisica e Scienze della Terra Università of Ferrara	Ferrara	Italy
Dr. CLOET, Ian	icloet@anl.gov	Argonne National Laboratory	Argonne	United States
Prof. CRABB, Donald	dgc3q@virginia.edu	University of Virginia	Charlottesville, VA22903	United States
Ms. CUI, Yimping	cuiyimpingwuli@163.com	Tsinghua University	Beijing	China
CUMMINGS, Melissa	melissac@jlab.org	The College of William and Mary	Williamsburg	United States
Dr. DHOSE, Nicole	nicole.dhose@cea.fr	CEA/IRFU,Centre d'etude de Saclay Gif-sur-Yvette (FR)	Gif-sur-Yvette F91191	France
Dr. DELTUVA, Arnoldas	deltuva@cii.fc.ul.pt	CFNUL	Lisboa	Portugal

Dr. DENG, Jian	jdeng@sdu.edu.cn	Shandong University	Jinan	China
Prof. DESHPANDE, Abhay	abhay.deshpande@stonybrook.edu	Stony Brook University	Stony Brook	United States
Dr. DIETER, Eversheim	d.eversheim@googlemail.com	Helmholtz Institut fuer Strahlen- und Kernphysik, University Bonn, Germany	Bonn	Germany
Mr. DILKS, Christopher	christopher.j.dilks@gmail.com	Pennsylvania State University	State College, PA	United States
DONG, Hui	hdong@sdu.edu.cn	Shandong University	Jinan	China
Prof. DUAN, ChunGui	duancg66@sina.com	Hebei normal University	Shijiazhuang	China
Mr. DUAN, zhe	duanz@ihep.ac.cn	Institute of High Energy Physics, Chinese Academy of Sciences	Beijing	China
Prof. DU, Jiangfeng	qcm@ustc.edu.cn	USTC	HEFEI	China
Prof. DUTTA, Dipangkar	d.dutta@msstate.edu	Mississippi State University	Mississippi State	United States
Ms. DU, Xiaozhen	duxiaozhenpku@gmail.com	Peking University	Beijing	China
Prof. EFREMOV, Anatoly	efremov@theor.jinr.ru	Joint Institute for Nuclear Research	Dubna	Russian Federation
Dr. ELLINGHAUS, Frank	ellingha@uni-mainz.de	Johannes-Gutenberg-Universitaet Mainz (DE)	Mainz	Germany
Dr. ENGELS, Ralf	r.w.engels@fz-juelich.de	Forschungszentrum Jülich	Jülich	Germany
ENT, Rolf	ceraul@jlab.org	Jefferson Lab	Newport News	United States
Prof. ERLER, Jens	erler@fisica.unam.mx	IF-UNAM	Mexico City	Mexico
Mr. FATEMI, Renee	renee.fatemi@uky.edu	University of Kentucky	Lexington, KY	United States
Dr. FERBER, Torben	torben.ferber@desy.de	DESY	Hamburg	Germany
Dr. FERRERO, Andrea	andrea.ferrero@cern.ch	CEA/IRFU,Centre d'etude de Saclay Gif-sur-Yvette (FR)	Gif-sur-Yvette	France
Prof. FIDECARO, Giuseppe	giuseppe.fidecaro@cern.ch	CERN	Geneva 23	Switzerland
FILIPPONE, brad	bradf@caltech.edu	caltech	pasadena	United States
Dr. FIMUSHKIN, Victor	fimushkin@jinr.ru	Joint Institute for Nuclear Research	Dubna	Russian Federation
Dr. FINGER, Michael	michael.finger@cern.ch	Charles University (CZ)	Prague	Czech Republic
Prof. FINGER, Miroslav	miroslav.finger@cern.ch	Charles University (CZ)	Prague	Czech Republic

Name	Email	Institution	City	Country
Prof. FISCHER, Horst	horst.fischer@cern.ch	Albert-Ludwigs-Universitaet Freiburg (DE)	Freiburg	Germany
Prof. FU, Changbo	cbfu@sjtu.edu.cn	Shanghai Jiaotong University	Shanghai	China
Dr. FURUKAWA, Takeshi	takeshi@tmu.ac.jp	Department of Physics, Tokyo Metropolitan University	Hachioji-shi, Tokyo 192-0397	Japan
Prof. GAO, Haiyan	gao@phy.duke.edu	Duke University	Durham, North Carolina	United States
GAO, Jian-Hua	gaojh@sdu.edu.cn	Shandong University	Weihai	China
Dr. GASPARYAN, Ashot	ashot.gasparyan@rub.de	Ruhr University of Bochum	Bochum	Germany
Dr. GIORDANO, Francesca	fgiord@illinois.edu	UIUC	Urbana-Champaign	United States
Mr. GONG, Ti	tigong@pku.edu.cn	Peking university	Beijing	China
GOTO, Yuji	goto@bnl.gov	RIKEN	Wako, Saitama	Japan
Mr. GOU, Boxing	boxinggou@gmail.com	Institute of Modern Physics, Chinese Academy of Sciences	Lanzhou	China
Prof. GROSSE-PERDEKAMP, Matthias	mgp@illinois.edu	Univ. Illinois at Urbana-Champaign (US)	Urbana, IL 61801	United States
Ms. GUAN, Yinghui	guanyh@hep.ac.cn	Institute of High Energy of Physics, CAS	Beijing	China
HARTMANN, Jan	hartmann@hiskp.uni-bonn.de	HISKP, Univerity of Bonn	Bonn	Germany
Dr. HATTA, Yoshitaka	hatta@yukawa.kyoto-u.ac.jp	Yukawa institute	Kyoto	Japan
Prof. HE, Han-Xin	hxhe@ciae.ac.cn	China Institute of Atomic Energy	Beijing	China
Prof. HE, Ke	kehe@mail.tsinghua.edu.cn	Tsinghua University	Beijing	China
HUANG, Haixin	huanghai@bnl.gov	BNL	Upton	United States
Mr. HUANG, Yanqi	yanqihuang@pku.edu.cn	Peking University	Beijing	China
Dr. JIANG, Xiaodong	jiang@jlab.org	Los Alamos National Laboratory	Los Alamos	United States
Mr. JI, wei	18701568519@163.com	Tsinghua University	Beijing	China
Prof. JI, xiangdong	xji@umd.edu	university of maryland	College Park	United States
Dr. KARYAN, Gevorg	gevkar@mail.desy.de	A. I. Alikhanyan National Science Laboratory	Yerevan	Armenia
Dr. KATCHARAVA, Andro	a.kacharava@fz-juelich.de	Nuclear Physics Institute (IKP)	Juelich	Germany

KAWALL, David	kawall@physics.umass.edu	University of Massachusetts Amherst	Amherst	United States
Mr. KAWASE, Shoichiro	kawase@cns.s.u-tokyo.ac.jp	Center for Nuclear Study, the University of Tokyo	Wako	Japan
Dr. KEITH, Christopher	ckeith@jlab.org	Jefferson Lab	Newport News	United States
KELLER, Dustin	dustin@jlab.org	University of Virginia	Charlottesville	United States
Prof. KISTRYN, Stanislaw	stanislaw.kistryn@uj.edu.pl	Jagiellonian University	Krakow	Poland
Dr. KOROTCHENKO, Konstantin	korotchenko@tpu.ru	National Research Tomsk Polytechnic University	Tomsk	Russian Federation
Prof. KOVALENKO, Alexander	kovalen@dubna.ru	Joint Institute for Nuclear Research	Dubna	Russian Federation
Prof. KRISCH, Alan	krisch@umich.edu	University of Michigan	Ann Arbor, Michigan 48109-1040	United States
Dr. KUBAROVSKY, Valery	val.kuba@gmail.com	Jefferson Laboratory	Newport News	United States
Prof. KUMANO, Shunzo	shunzo.kumano@kek.jp	KEK	Tsukuba	Japan
Prof. KUMAR, Krishna	kkumar@physics.umass.edu	Stony Brook University	Stony Brook, NY	United States
Dr. KUMERICKI, Kresimir	kkumer@phy.hr	University of Zagreb	Zagreb	Croatia
Dr. KUNNE, Fabienne	fabienne.kunne@cern.ch	CEA/IRFU,Centre d'etude de Saclay Gif-sur-Yvette (FR)	Gif-sur-Yvette	France
Prof. KUREK, Krzysztof	kurek@fuw.edu.pl	National Centre for Nuclear Research (PL)	Warsaw	Poland
Prof. LADYGIN, Vladimir	vladygin@jinr.ru	LHEP-JINR	Dubna	Russian Federation
Dr. LANSBERG, Jean-Philippe	lansberg@in2p3.fr	IPN Orsay, Paris Sud U. / IN2P3-CNRS	Orsay	France
Prof. LEHRACH, Andreas	a.lehrach@fz-juelich.de	Forschungszentrum Juelich	52428 Juelich	Germany
Prof. LENISA, Paolo	lenisa@fe.infn.it	University of Ferrara and INFN	FERRARA	Italy
Dr. LIANG, Haozhao	haozhao.liang@riken.jp	RIKEN	Wako	Japan
Prof. LIANG, Zuo-tang	liang@sdu.edu.cn	Shandong University	Jinan	China
Prof. LI, Liang	liangliphy@sjtu.edu.cn	Shanghai Jiao Tong University (CN)	Shanghai	China
Dr. LIN, Fanglei	fanglei@jlab.org	Thomas Jefferson National Accelerator Facility	Newport News	United States

Name	Email	Institution	City	Country
Prof. LIU, Keh-Fei	liu@pa.uky.edu	University of Kentucky	Lexington, KY	United States
Dr. LIU, Ming	mingxiong.liu@gmail.com	Los Alamos National Laboratory	Los Alamos	United States
Mr. LIU, Tianbo	liutb@pku.edu.cn	Peking University	Beijing	China
Dr. LIU, Zheng-Xin	liuzxqh@tsinghua.edu.cn	Institute for Advanced Study, Tsinghua University	Beijing	China
LI, Xiaomei	xiao_mei_li@yahoo.com	China Institute of Atomic Energy (CN)	Beijing	China
Prof. LONG, Gui Lu	gllong@tsinghua.edu.cn	T	Beijing	China
Prof. LORENZON, Wolfgang	lorenzon@umich.edu	Michigan	Ann Arbor	United States
Prof. LU, Li	junren2006@gmail.com	Institute of Physics, Chinese Academy of Science	Beijing	China
LU, Zhun	zhunlu@seu.edu.cn	Southeast University	Nanjing	China
Prof. MAAS, Frank	maas@uni-mainz.de	Helmholtz Institute Mainz	55128 Mainz	Germany
Prof. MA, Bo-Qiang	mabq@pku.edu.cn	Peking University	Beijing	China
Dr. MACHARASHVILI, Giorgi	gogi@nusun.jinr.ru	JINR	Dubna, 141980	Russian Federation
Dr. MACK, David	mack@jlab.org	TJNAF	Newport News	United States
Dr. MAKDISI, Yousef	makdisi@bnl.gov	Brookhaven National Laboratory	Upton, NY 11973	United States
Dr. MAKKE, Nour	nour.makke@cern.ch	Universita e INFN (IT)	Trieste	Italy
Dr. MALLOT, Gerhard	gerhard.mallot@cern.ch	CERN	Geneva	Switzerland
Dr. MANION, Andrew	manionan@gmail.com	PHENIX Experiment	Lake Ronkonkoma	United States
Dr. MAO, Wenjuan	wjmao@seu.edu.cn	Room North 504 of Physics Building of Southeast University, Nanjing 211189, China	Nanjing	China
MARTIN, Anna	anna.martin@cern.ch	Trieste University and INFN	Trieste	Italy
Dr. MARUKYAN, Hrachya	marukyan@mail.desy.de	A.I. Alikhanian National Science Laboratory	Yerevan	Armenia
Prof. MASAIKE, Akira	masaike@mvb.biglobe.ne.jp	Kyoto University	Abiko-shi, Chiba-ken	Japan
Dr. MASSACRIER, Laure Marie	laure.marie.massacrier@cern.ch	LAL/IPNO	Orsay	France
Dr. MAXWELL, James	jdmax@mit.edu	MIT	Cambridge, MA	United States

Name	Email	Institution	City	Country
Dr. MA, Yan-Qing	yqma.cn@gmail.com	BNL	Upton	United States
Dr. MA, Yue	y.ma@riken.jp	RIKEN	Saitama	Japan
Mr. MEI, Jincheng	mumeijc@gmail.com	Shandong University	Jinan	China
Prof. MENG, Jie	mengj@pku.edu.cn	Peking University	Beijing	China
Prof. MEYER, Werner Peter	werner.meyer@cern.ch	Ruhr-Universitaet Bochum (DE)	Bochum	Germany
MEY, Sebastian	s.mey@fz-juelich.de	Forschungszentrum Jülich	53121 Bonn	Germany
Prof. MEZIANI, Zein-Eddine	meziani@temple.edu	Temple University	Philadelphia	United States
Mr. MEZRAG, Cedric	cedric.mezrag@cea.fr	IRFU/SPhN	Gif-Sur-Yvette	France
Prof. MIBE, Tsutomu	mibe@post.kek.jp	IPNS, KEK	Ibaraki	Japan
Prof. MILNER, Richard	milner@mit.edu	MIT	Cambridge, MA 02139	United States
Prof. MISKIMEN, Rory	miskimen@physics.umass.edu	University of Massachusetts	Amherst	United States
Dr. MOCHALOV, Vassili	mochalov@ihep.ru	IHEP	Protvino	Russian Federation
Dr. MOCHALOV, Vassili	vassili.mochalov@cern.ch	Institute for High Energy Physics (RU)	Protvino	Russian Federation
Dr. MOROZOV, Vasiliy	morozov@jlab.org	Thomas Jefferson National Accelerator Facility	Newport News	United States
Dr. MUELLER, Dieter	dieter.mueller@tp2.rub.de	Ruhr-University Bochum	Bochum	Germany
MULDERS, Piet	p.j.g.mulders@vu.nl	VU/Nikhef	Amsterdam	Netherlands
Dr. NAGAYTSEV, Alexander	alexander.nagaytsev@cern.ch	Joint Inst. for Nuclear Research (RU)	Dubna	Russian Federation
Mr. NAIR, Sreeraj	sreeraj_nair@iitb.ac.in	Indian Institute of Technology, Bombay	Mumbai	India
Dr. NAKANO, Kenichi	knakano@nucl.phys.titech.ac.jp	Tokyo Tech	Tokyo	Japan
Ms. NIE, Pin	turelong@gmail.com	huazhong university of science and technology	wuhan	China
Prof. NI, Wei-Tou	weitou@gmail.com	National Tsing Hua University	Hsinchu	Taiwan
NOCERA, Emanuele Roberto	emanuele.nocera@unimi.it	Università degli Studi di Milano & INFN Milano,Italy	Milan	Italy
OGAWA, akio	akio@bnl.gov	BNL	Upton	United States

Name	Email	Institution	City	Country
OU, Li	liou@gxnu.edu.cn	Guangxi Normal University	Guilin	China
Prof. PAN, Xinyu	xypan@aphy.iphy.ac.cn	Institute of Physics, Chinese Academy of Sciences	Beijing	China
Mr. PAN, Yuxi	yuxipan@physics.ucla.edu	University of California, Los Angeles	Los Angeles	United States
Dr. PARSAMYAN, Bakur	bakur.parsamyan@cern.ch	INFN and University of Turin	Turin	Italy
Dr. PASQUINI, Barbara	pasquini@pv.infn.it	University of Pavia	Pavia	Italy
PASYUK, Eugene	pasyuk@jlab.org	Jefferson Lab	Newport News, Virginia 23606	United States
Dr. PESHEKHONOV, Dmitri	dimitri.pechekhonov@cern.ch	Joint Inst. for Nuclear Research (RU)	Dubna	Russian Federation
Dr. PING, Ronggang	pingrg@ihep.ac.cn	IHEP	Beijing	China
PISKUNOV, Nikolay	piskunov@jinr.ru	Joint Institute for Nuclear Research	Dubna, Moscow region 141980	Russian Federation
Dr. POELKER, Matthew	poelker@jlab.org	Jefferson Lab	Newport News	United States
Prof. PREPOST, Richard	prepost@hep.wisc.edu	University of Wisconsin	Madison Wisconsin	United States
PROKUDIN, Alexei	prokudin@jlab.org	Jefferson Lab	Newport News	United States
Dr. PTITSYN, Vadim	vadimp@bnl.gov	Brookhaven National Laboratory	Upton	United States
QIU, Jianwei	jqiu@bnl.gov	Brookhaven National Lab	Upton, NY 11973	United States
Dr. RANJBAR, Vahid	vranjbar@bnl.gov	BNL	Upton	United States
Dr. REICHERZ, Gerhard Alois	gerhard.reicherz@rub.de	Ruhr-University Bochum	Witten	Germany
Dr. REICHERZ, Gerhard Alois	reicherz@tau.ep1.rub.de	Ruhr-Universitaet Bochum (DE)	Bochum	Germany
Dr. ROSER, Thomas	roser@bnl.gov	BNL	Upton, NY 11973-5000	United States
Dr. ROSTOMYAN, Ami	ami@mail.desy.de	DESY	Hamburg	Germany
Prof. SAGAWA, Hiroyuki	hiroyuki.sagawa@gmail.com	RIKEN Nishina Center	Wako	Japan
Prof. SAITO, Naohito	naohito.saito@kek.jp	KEK / J-PARC	Tsukuba	Japan
Dr. SAKAGUCHI, Satoshi	sakaguchi@phys.kyushu-u.ac.jp	Kyushu University	Fukuoka	Japan
Mr. SALEEV, Artem	a.saleev@fz-juelich.de	Samara State University	Samara	Russian Federation
Dr. SANDORFI, Andrew	sandorfi@jlab.org	Thomas Jefferson National Accelerator Laboratory	Newport News	United States

Mr. SBRIZZAI, Giulio	giulio.sbrizzai@cern.ch	Universita e INFN (IT)	trieste	Italy
Prof. SEKIGUCHI, Kimiko	kimiko@lambda.phys.tohoku.ac.jp	Tohoku University	Sendai	Japan
Dr. SEMENOV, Pavel	pavel.semenov@ihep.ru	IHEP	Protvino	Russian Federation
Dr. SHAN, Pujia	shanpujia@126.com	ICQM	Beijing	China
Dr. SHARMA, Neetika	neetika@iisermohali.ac.in	Indian Institute of Science Education and Research Mohali	Mohali-140306 Punjab	India
Prof. SHATUNOV, Yury	shatunov@inp.nsk.su	budker institute of nuclear physics	Novosibirsk	Russian Federation
Dr. SHKLYAR, Vitaly	shklyar@theo.physik.uni-giessen.de	University of Giessen	Giessen	Germany
SICHTERMANN, Ernst	epsichtermann@lbl.gov	Lawrence Berkeley National Laboratory	Berkeley	United States
Dr. SKIBINSKI, Roman	roman.skibinski@uj.edu.pl	Jagiellonian University	Krakow	Poland
Dr. SKOBY, Michael	mjskoby@gmail.com	Indiana University	Bloomington	United States
Prof. SNOW, William	wsnow@indiana.edu	Indiana University	Bloomington	United States
Dr. SONG, Yu-kun	songyk@ustc.edu.cn	University of Science and Technology of China	Hefei	China
SOUDER, Paul	souder@physics.syr.edu	Syracuse University	Syracuse	United States
Prof. STEFFENS, Erhard	steffens@physik.uni-erlangen.de	Dept. of Physics, University of Erlangen-Nürnberg	Erlangen	Germany
Dr. STEPHENSON, Edward	stephene@indiana.edu	Indiana University	Bloomington 47408	United States
STOLARSKI, Marcin	marcin.stolarski@cern.ch	LIP Laboratorio de Instrumentacao e Fisica Experimental de Part	Lisboa	Portugal
Prof. STROEHER, Hans	h.stroeher@fz-juelich.de	Forschungszentrum Juelich GmbH	Juelich	Germany
Dr. SUN, peng	psun@lbl.gov	L	Berkeley	United States
Dr. SVIRIDA, Dmitry	dmitry.svirida@itep.ru	ITEP	Moscow	Russian Federation
Mr. TAO, Lei	1245401524@qq.com	Huazhong University of Science & Technology	Wuhan	China
Mr. TER WOLBEEK, Johannes	jwolbeek@cern.ch	Albert-Ludwigs-Universitaet Freiburg (DE)	Freiburg	Germany
Prof. TERYAEV, Oleg	teryaev@theor.jinr.ru	JINR	Dubna	Russian Federation
Dr. THOMAS, Andreas	thomas@kph.uni-mainz.de	University Mainz	Mainz	Germany

Name	Affiliation	Email	City	Country
Dr. TOPORKOV, Dmitriy	Budker Institute of Nuclear Physics	d.k.toporkov@inp.nsk.su	Novosibirsk	Russian Federation
Dr. TORII, Hiroyuki	University of Tokyo (JP)	torii@radphys4.c.u-tokyo.ac.jp	Tokyo	Japan
Dr. TSENTALOVICH, Evgeni	MIT-Bates	evgeni@mit.edu	Middleton	United States
Dr. TULLNEY, Kathlynne	University of Mainz	tullnek@uni-mainz.de	Mainz	Germany
UENO, Hideki	RIKEN Nishina Center for Accelerator-Based Science	ueno@riken.jp	Saitama	Japan
Mr. VAN DER VEKEN, Frederik	University of Antwerp	frederik.vanderveken@ua.ac.be	Antwerp	Belgium
Dr. VAUTH, Annika	Deutsches Elektronen-Synchrotron (DE)	annika.vauth@desy.de	Hamburg	Germany
Mr. WADA, Yasunori	Tohoku University	wada@lambda.phys.tohoku.ac.jp	Sendai	Japan
Mr. WANG, Bin	Huazhong University of Science & Technology	wangbinphys@hust.edu.cn	Wuhan	China
Dr. WANG, LI	Donghua University	wang_li@dhu.edu.cn	Shanghai	China
Dr. WANG, Rong	Institute of Modern Physics, Chinese Academy of Sciences	rwang@impcas.ac.cn	Lanzhou	China
Dr. WANG, Xiaorong	New Mexico State University and RBRC	xwang@nmsu.edu	Las Cruces	United States
WANG, Yunxiao	University of Virginia	yw6vp@virginia.edu	Charlottesville	United States
Mr. WANG, Zhen-Lai	Huazhong University of Science & Technology	wangzhenlai@hust.edu.cn	Wuhan	China
Mr. WANG, Zhiguo	National University of Defense Technology	maxborn@163.com	Changsha	China
WARREN, Warren	Duke University	warren.warren@duke.edu	Durham	United States
WEI, Shu-yi	Shandong University	spinphysics.hep@gmail.com	Jinan	China
Dr. WEI, Xiangdong	Jefferson Lab	xwei@lab.org	Newport News	United States
Mr. WUNDERLICH, Yannick	University of Bonn	wunderlich@hiskp.uni-bonn.de	Bonn	Germany
Prof. XIAO, Bowen	Central China Normal University	bowen@phys.columbia.edu	Wuhan	China
Prof. XIAO, Zhigang	Tsinghua University, Department of Physics	xiaozg@tsinghua.edu.cn	Beijing	China
Prof. XU, Qinghua	Shandong University	xuqh@sdu.edu.cn	Jinan	China

Mr. XU, Zhaojie	motoch@aliyun.com	Shanghai Jiao Tong University	Shanghai	China
Mr. YANG, Detian	530188229@qq.com	Huazhong University of Science & Technology	Wuhan	China
Dr. YANJUN, Sun	sunyanjun@nwnu.edu.cn	Northwest Normal University	Lanzhou	China
Mr. YASUDA, Jumpei	yasuda@phys.kyushu-u.ac.jp	Department of Physics, Kyushu University	Fukuoka	Japan
Dr. YOSHIDA, Shinsuke	shinsuke.yoshida@riken.jp	Brookhaven National Laboratory	New York	United States
Prof. YU, Haiming	haiming_yu@buaa.edu.cn	Beihang University	Beijing	China
Mr. YU, Haiwang	yuhw.pku@gmail.com	Peking University	Beijing	China
Dr. YURIY, uzikov	uzikov@jinr.ru	Joint Institute for Nuclear Researches	Dubna	Russian Federation
Mr. YUXIANG, zhao	yxzhao@mail.ustc.edu.cn	USTC	Hefei	China
Dr. ZANOTTI, James	james.zanotti@adelaide.edu.au	University of Adelaide	Adelaide	Australia
Dr. ZELENSKI, Anatoli	zelenski@bnl.gov	BNL	Upton	United States
Dr. ZENIHIRO, Juzo	juzo@ribf.riken.jp	RIKEN Nishina Center	Saitama	Japan
Mr. ZHANG, Baiyang	astris.dei@gmail.com	Institute of Modern Physics,CAS	Lanzhou	China
Dr. ZHANG, Jian-Rong	jrzhang@nudt.edu.cn	National University of Defense Technology	Changsha	China
Dr. ZHANG, Jianhui	zhangjianhui@gmail.com	SJTU	Shanghai	China
Mr. ZHANG, Jinlong	jlzhang@rcf.rhic.bnl.gov	Shandong University/Brookhaven National Laboratory	Upton	United States
Dr. ZHANG, Pengming	zhpm@impcas.ac.cn	Institute of Modern Physics, Lanzhou, China	Lanzhou	China
Prof. ZHANG, Tong	tzhang18@fudan.edu.cn	Fudan University	Shanghai	China
Dr. ZHANG, Zhenyu	zhenyuzhang@whu.edu.cn	Wuhan university	Wuhan	China
Mr. ZHAO, Yong	yongzhao@umd.edu	University of Maryland, College Park	College Park	United States
Dr. ZHOU, Xiang	xiangzhou@whu.edu.cn	Wuhan University	China	China
Dr. ZHOU, Yajin	zhouyj@sdu.edu.cn	Shandong Univ.	Jinan	China
Dr. ZHOU, jian	zzzhoujian@gmail.com	Regensburg University	Regensburg	Germany

Dr. ZIELINSKI, Marcin   m.zielinski@uj.edu.pl   Jagiellonian University   Krakow   Poland

Mr. ZIWEI, Chen   405029537@qq.com   HUST   Wuhan,Hubei   China

Printed in the United States
By Bookmasters